Medicine in Homer

Medicine in Homer

Edited by

Eric L. Altschuler

Essen Health Care, Bronx, NY, United States;
Kessler Institute for Rehabilitation,
West Orange, NJ, United States

ELSEVIER

ACADEMIC PRESS

An imprint of Elsevier

Academic Press is an imprint of Elsevier
125 London Wall, London EC2Y 5AS, United Kingdom
50 Hampshire Street, 5th Floor, Cambridge, MA 02139, United States

ISBN: 978-0-443-32978-4

For Information on all Academic Press publications
visit our website at https://www.elsevier.com/books-and-journals

Publisher: Stacy Masucci
Acquisitions Editor: Megan McManus
Editorial Project Manager: Kyle Gravel
Production Project Manager: M. Balakrishnan
Cover Designer: Vicky Pearson Esser

Typeset by MPS Limited, Chennai, India

Dedication

To our heroes and heroines past, present, and future and their physicians and healthcare providers.*

* The image on the cover shows one panel from the amazing vase known as the Euphronios Krater (c. 515 BCE) in honor of the artist. The panel depicts the injuries to staunch and mighty Trojan ally Sarpedon, son of Zeus from Lycia. See Note 3 in Chapter 10 on page 104 for further discussion.

Contents

List of contributors

Eric L. Altschuler Essen Health Care, Bronx, NY, United States; Kessler Institute for Rehabilitation, West Orange, NJ, United States

Fabian Horn Ludwig-Maximilians-University Munich, Department of Greek and Latin Philology, Munich, Germany

Christos Koutserimpas School of Rehabilitation Health Sciences, University of Patras, Greece

Anusha Lekshminarayanan Department of Rehabilitation Medicine, New York Medical College, Valhalla, NY, United States; Department of Physical Medicine and Rehabilitation, Westchester Medical Center, Valhalla, NY, United States

Tyler Lucas Division of Orthopaedic Surgery, Department of Surgery, Metropolitan Hospital, New York, NY, United States

Pádraig MacCarron Department of Mathematics & Statistics, University of Limerick, Limerick, Ireland

Mehmet Oguz Namal Bern, Switzerland

Mark Pagel University of Reading, Reading, United Kingdom

Gregory T. Para Sarasota Parrot Conservatory, Sarasota, FL, United States

Andreas Plaitakis University of Crete, Heraklion, Greece; Icahn School of Medicine at Mount Sinai, New York, NY, United States

Simone Swartz Author and Service Dog Strong Founder, Portland, ME, United States

Acknowledgments

I thank my teachers—Mrs. Barbara Ellis (Latin), Dr. Jamo Blake, PhD (Latin, Homeric Greek), Mr. Abraham Baumel (Principal) and Murray Kahn (Assistant Principal) (Latin and Greek contest administration), Prof. Hayden Pelliccia (Attic Greek), Prof. Calvert Watkins (Hittite) and Dr. Nicholas Halasz, MD (anatomy, surgery).

Thanks to my fantastic colleagues from whom you will hear in the chapters of the book. These include one student, now a colleague (Anusha Lekshminarayanan, MD). Thanks to Megan McManus, Kyle Gravel, Christine McElvenny, Selvaraj Raviraj, and the entire incredible Editorial, ELSA, and Production teams from Elsevier for bringing to fruition a book that brings to life the anatomy and trauma treatment and management in Homer's *Iliad*.

I thank Mileidy Fernandez whom I found on Fiverr for the amazing anatomic figures in the book. Homer has never been better illustrated. I thank Sheila Brody for helpful discussions.

Eric L. Altschuler

Section 1

Introduction

Introduction, synopsis of chapters, and the Homeric question

1

Eric L. Altschuler[1,2]
[1]Essen Health Care, Bronx, NY, United States, [2]Kessler Institute for Rehabilitation, West Orange, NJ, United States

Who was Homer? That's an easy question. He wrote the *Iliad* and *Odyssey*.[1] Others say that the same person did not write both epics. On this point, I agree strongly with one of Homer's great English translators Richard Lattimore who said in the introduction to his translation of the *Odyssey* (Homer, 2007) that either Homer also wrote the *Odyssey* or someone who had studied and knew the *Iliad* as intently and intimately as its author wrote the *Odyssey*. (For more on this see Chapter 21).

What else do we know about Homer? Nothing. Yes. Nothing. We don't know where Homer was from, when he lived, or even what dialect of Greek he wrote in. We don't know anything about Homer's background or how he lived and what work he might have done other than writing the two epics. Similarly, we don't know if the story of the siege of Troy in the *Iliad* is an account of real happenings and we know nothing of Homer's working methods.

The situation is worse than for Shakespeare as there is no literature or historical written sources to which to compare Homer. Archeology is not helpful as we don't know when Homer lived—with proposed dates ranging from 800 to 1200 BCE.

Ignorance about all these issues and along all these fronts comprise what is known as the Homeric question. (We might further ask how Homer supported himself while he was writing the epics.)

It seems hopeless to think that we can learn anything about Homer. But we *do* have something to which to compare Homer. Human anatomy! Human anatomy has not changed in the last 2800 years or 3200 years. First cataloged in the 19th century by Fröhlich (Fröhlich, 1879) there are descriptions of more than 140 battle injuries in the *Iliad*. This has even led to suggestions that Homer was a physician, indeed a surgeon (Godquin, 1990; Melchionda, 1963; Melchionda, 1964).

We can go further and try to surmise what level of experience as surgeon Homer might have been: Homer not only shows an interest in battle injuries but in doctors as well. As discussed in Chapter 3 quite a number are mentioned in the *Iliad*. Two in particular stand out: Machaon, a surgeon and his brother Podalirius a physician. Their father is the legendary Asclepius—he of the caduceus—who himself was taught medicine by centaur Chiron.

Machaon is the Chief Surgeon among the Greek forces and Podalirius the Chief Physician. Chiron also taught some medical skills to the great warrior king Achilles

[1] Some people think they were written by a committee. Obviously proponents of this theory have never been on a committee that issued a report.

Medicine in Homer. DOI: https://doi.org/10.1016/B978-0-443-32978-4.00006-7

(and Achilles taught his comrade Patroclus). This then places Asclepius as the "Founding Dean" of Greek medicine. So Homer was interested not only in injuries and medicine, but also, somehow, *organized medicine*—centuries before its known formal organization. Homer's respect without jealousy or disillusionment of the medical hierarchy suggests he was a senior surgical trainee or new junior qualified surgeon. The idea that Homer may have been a surgeon would put an interesting personal twist on the episode in Book 11 of the *Iliad* when Machaon is injured and the wise old man Nestor, King of Pylos, is called in to help care for Machaon. (Nestor seems to have been quite literally "grandfathered" into the medical guild when they established the Board.)

Examination of Homer's description of injuries, independent of the idea that Homer may have been a surgeon, also yields insight into his poetic, rhetorical style: Homer rarely mixes metaphors of real injuries with metaphoric concepts on heart, lungs, and mind, whereas subtle combinations and transmutations of these are the hallmark and indeed practically define Shakespeare's style.

Cataloging, describing, characterizing, and analyzing the injuries in the *Iliad* and the *Odyssey* leads one then naturally to consider the hypothesis that Homer was a junior surgeon who had been embedded with a Greek army on some war campaign.

In Chapter 2, I review Alwyn W Beasley MD's informative, intriguing, inspiring and entertaining article (Beasley, 1972) about orthopedics in Homer. This is the article that got me interested in medicine in Homer and is really the foundational article on the topic in the Twentieth century.

In Chapter 3, orthopedic surgeon Christos Kousterimpas briefly introduces, surveys and discusses the physicians and many other medical personnel in the *Iliad* and the *Odyssey*.

In Chapter 4, I discuss the curious character Thersites, the only character in the *Iliad* without a patronymic or country, his medical condition, and relation of this to Homer and the epics. In Chapter 5, I make suggestions for which translations of Homer to use.

The second section **Trauma in Homer** is the heart of the book. In Chapter 6, orthopedic surgeon Christos Kousterimpas gives an updated (Fröhlich, 1879) trauma registry for the *Iliad*. In Chapter 7, Koutserimpas supplies the trauma registry for Homer's *Odyssey*, and in Chapter 8, he describes and discusses trauma medicine and treatment in Homer. In Chapters 9–13, Tyler Lucas, MD, Chief of Orthopaedic Surgery at Metropolitan Hospital, Anusha Lekshminarayanan, MD, a Sports Medicine and Rehabilitation Medicine doctor from Westchester Medical Center and myself catalog and analyze the injuries to the upper extremity (Chapter 9), thorax (Chapter 10), abdomen (Chapter 11) and lower extremity and pelvis (Chapter 12) in the *Iliad*. In Chapter 13, the extremely lethality of injuries to the head and neck in the *Iliad* is noted and cataloged. In Chapter 14, Dr. Habil. Fabian Horn, M.A. from the Department of Greek and Latin Philol at Ludwig-Maximilians-Universität München discusses trauma injuries in Homer from a classicist's point of view.

The third section **Neurology, Toxicology, Genetics and Infectious Disease in Homer** covers nontrauma aspects of medicine in Homer. In the first chapter in the section (Chapter 15) Anusha Lekshminarayanan, and I discuss exceptional cases of neurologic disease and neurotrauma in Homer. In Chapter 16, I note and describe a case of antisocial personality disorder (ASPD) in Homer. In Chapter 17, physician and biologist Andreas Plaitakis discusses plants, poisons, potions, and plagues in the Homeric Epics. Chapter 18 isn't exactly modern genetics but composer and electronic music specialist Mehmet Oğuz Namal and I describe the eight-generation genealogy Homer gives for the Trojan royal family. (Oğuz also made the Bach musical score examples in Chapter 21.)

The fourth and final section of the book **Further Perspectives on Medicine in Homer** dissects the *Iliad* (and *Odyssey*) from points of view other than anatomy or medicine to further better understand and appreciate the Epics and Homer:

In Chapter 19, Professor Mark Pagel, FRS from the University of Reading uses genetic methods to try to date the composition of the *Iliad* and *Odyssey*. In Chapter 20 Pádraig MacCarron from the University of Limerick examines the social ("Facebook") networks of characters in the *Iliad* and the *Odyssey*. This is a fascinating approach to using the text of the epics to provide insight into the stories and possibly Homer's approach to their composition. In Chapter 21, I explore the question of whether the same writer wrote both the *Iliad* and the *Odyssey*.

Along with human anatomy another thing which hasn't changed in 2800 years or 3200 years is women, men and their relationships. In Chapter 22 army veteran and PTSD survivor Simone Swartzentruber-Emmons examines women, men, and posttraumatic stress disorder (PTSD) in Homer's *Iliad* with many surprising and powerful observations, analyses, and conclusions.

In the penultimate chapter (Chapter 23) combat veteran Greg Para "red teams" the hypothesis that Homer was a surgeon deployed with a Greek army vs. just having heard about war while sitting at some wine bar in Crete.

In the final chapter (Chapter 24), I summarize the findings in the book and suggest topics and directions for future research.

Notes:

1. Technical linguistic terms will be printed in ALL CAPS and then defined.

Table 1.1 Conventions for numbering the books of the *Iliad* and Odyssey.

Iliad book (Arabic numerals)	Greek letters	*Odyssey* book	Greek letters	Name of Greek letter
1	Α	1	α	alpha
2	Β	2	β	beta
3	Γ	3	γ	gamma
4	Δ	4	δ	delta
5	Ε	5	ε	epsilon
6	Ζ	6	ζ	zeta
7	Η	7	η	eta
8	Θ	8	θ	theta

(*Continued*)

Table 1.1 (Continued)

Iliad book (Arabic numerals)	Greek letters	*Odyssey* book	Greek letters	Name of Greek letter
9	Ι	9	ι	iota
10	Κ	10	κ	kappa
11	Λ	11	λ	lambda
12	Μ	12	μ	mu
13	Ν	13	ν	nu
14	Ξ	14	ξ	xi
15	Ο	15	ο	omicron
16	Π	16	π	pi
17	Ρ	17	ρ	rho
18	Σ	18	σ	sigma
19	Τ	19	τ	tau
20	Υ	20	υ	upsilon
21	Φ	21	φ	phi
22	Χ	22	χ	chi
23	Ψ	23	ψ	psi
24	Ω	24	ω	omega

2. The twenty-four books of the *Iliad* (and *Odyssey*) traditionally have been denoted by the twenty-four letters of the Greek alphabet with upper case letters for the books of the *Iliad* (and lowercase letters for books of the *Odyssey*).[2] Some authors in this book use Greek letters to number the books of the *Iliad* and the *Odyssey* while others use Arabic numbers. These two conventions are displayed in Table 1.1. [Also, this provides a good start and impetus to learning the Greek alphabet and Homeric Greek (see Chapter 5)!]

References

Beasley, A. W. (1972). Homer and orthopaedics. *Clinical Orthopaedics and Related Research, 89*, 10–16.

Fröhlich, H. (1879). *Die Militär Medizin Homer's*. Enke.

Godquin, B. (1990). [Was Homer a surgeon?]. *Chirurgie; memoires de l'Academie de chirurgie, 116*(2), 136–143.

Homer (2007). *The Odyssey of Homer*. HarperCollins.

Melchionda, E. (1963). [Homer, a military physician? (I: The *Iliad*)]. *Annali di Medicina Navale, 68*.

Melchionda, E. (1964). [Homer, a military physician? (I: The *Iliad*)]. *Annali di Medicina Navale, 69*.

[2] See Chapter 5 for discussion of the ghostly traces of a 25th letter of the alphabet.

Homer and orthopedics

2

Eric L. Altschuler[1,2]
[1]Essen Health Care, Bronx, NY, United States, [2]Kessler Institute for Rehabilitation, West Orange, NJ, United States

In one class in my first 2 years in medical school (the "pre-clinical years"), pharmacology I believe, the Professor taught us about a general theory a doctor/scientist in the 1700s (I think) had for the treatment of disease known as the Place Theory. The idea was that to treat or cure a disease all one needed to do was to find out where the disease had originated. How would this cure a disease? Well, the proposer of the theory said that when you figured out where the disease originated, a-go to that place, b-you would find that animals from the locale did not have the disease, c-then you found out what local plants the animals were eating to stave off the disease and d-you would have your treatment.

After hearing about this, being interested myself in finding causes and treatments of diseases and also being interested in history I thought I could perhaps there might be a *time* version of the theory: if one could figure out *when* a disease first originated this could be a crucial clue to finding its cause and perhaps ultimately a preventative measure, treatment, or cure.

A recent example of the successful use of temporal origin to find the cause of a disease—though with a number of dead ends—was the search for the cause of nephrogenic systemic fibrosis (NSF). Patients with NSF, first described in 1997 and originally known as nephrogenic systemic dermatopathy (Cowper et al., 2000; Cowper, 2003), suffer from fibrosis of the skin, joints, eyes, and internal organs. NSF was a new disease appearing in the human population only at the end of the Twentieth century. So its cause most likely must also be new as well. While being on hemodialysis was known to be a risk factor, the cause of NSF was not known (Moschella et al., 2004). Ultimately, the temporal origin clue paired with good old-fashioned biopsies of skin lesions of patients with NSF cracked the case: Biopsies of the lesions (Grobner, 2006; Marckmann et al., 2006; Centers for Disease Control and Prevention (CDC), 2007) were found to contain gadolinium, a dye used in MRI studies to increase contrast. This was the cause of the disease! The disease did not exist earlier, because before dialysis and MRI patients did not get the contrast or have so little kidney function left to process it. It had been known that iodine used for contrast in CT ("cat") scans can cause kidney failure (and even pulmonary edema) when administered to patients with renal disease, but this had not been a known problem for MRI contrast. Now gadolinium is not used in dialysis patients and NSF has largely been eliminated.

Data from more ancient times can also be helpful. Indeed, while coronary artery disease is often thought to be the product of the diet and lifestyle of modernity,

Medicine in Homer. DOI: https://doi.org/10.1016/B978-0-443-32978-4.00019-5

Thompson et al. (2013) studied mummies from four different preindustrial civilizations scattered around the globe and found CAD in a nontrivial percentage in all of them suggesting that there may be a basic human predisposition to the condition.

As far as psychiatric disease the occurrence of antisocial personality disorder (ASPD) in the Biblical character Samson son of Manoah (Altschuler et al., 2001) confirms that whatever the cause for ASPD, it must have been present for at least 3000 years in humans, if not far longer. (See also Chapter 16 for a case of ASPD in Homer.) Conversely, by searching historical and literary texts and asylum log books, it seems that somehow schizophrenia may be a disease that first appeared only ~1800 (see Altschuler (2001) and refs. therein).

So history and literature can be a guide to disease causes and possibly even suggest potential treatments. But we can also turn this thinking around and use disease appearance in literature to perhaps tell us something about literary authors—in the very least that *they observed a given disease.*

Having studied classical languages (Latin and Greek) since middle school I wondered if medicine might have something to teach us, then, about Homer. I was thus intrigued when I came across an article about orthopedics in Homer (Beasley, 1972) by Alwyn W Beasley, MD an orthopedic surgeon from New Zealand.

The article (Beasley, 1972) is an entertaining, eclectic, and stimulating romp through orthopedic cases in Homer. Beasley's article is a founding paper for the twentieth and twenty-first centuries of medicine in Homer and also what sparked my interest in the topic. Beasley discusses descriptions of three groups of orthopedic cases in Homer: (1) physical descriptions by Homer permitting the diagnosis of a congenital deformity, (2) descriptions of wounds, and (3) descriptions of treatments provided.

Under deformities, Beasley lists the case of Hephaestus the Greek god of artisans, blacksmiths, and other craftsmen. Hephaestus is the son of Hera. Based on (1) comments in the *Iliad* such as (*Iliad* Book 21, lines 300–301) And forthwith she [Hera] spake unto Hephaestus, her dear son: "Rouse thee, Crook-foot, my child!", (2) descriptions of Hephaestus, e.g., (*Iliad* Book 18, lines 410–411) He [Hephaestus] spake, and from the anvil rose, a huge, panting bulk, halting the while, but beneath him his slender legs moved nimbly, Beasley diagnoses Hephaestus with bilateral congenital club foot. [Translations here and following from (Homer, 1924) licensed under a Creative Commons Attribution-ShareAlike 3.0 United States License.]

Beasley next lists examples of orthopedic wounds/injuries in the *Iliad*. These will be covered in detail in Chapters 9, 10 and 12. Beasley notes that Homer had a not insignificant knowledge of orthopedic anatomy and an eye for detail.

In particular, Beasley notes that Homer makes quite a number of references to the collar bone: (*Iliad* Book 5, lines 146–148) and the other he struck with his great sword upon the collar-bone beside the shoulder, and shore off the shoulder from the neck and from the back; (*Iliad* Book 17, lines 309–312) Him [Schedius, son of Iphitus] Hector smote beneath the midst of the collar-bone, and clean through passed the point of bronze, and came out beneath the base of the shoulder. And he fell with a thud, and upon him, his armor clanged; (*Iliad* Book 8, lines 320–329) And himself Hector leapt to the ground from his gleaming car crying a terrible cry, and seizing a stone in his hand made right at Teucer, and his heart bade him smite him. Now Teucer had drawn forth from the quiver a bitter arrow, and laid it upon the string, but even as he was drawing it back Hector of the flashing helm smote him beside the shoulder where the collar-bone parts the neck and the breast, where is the deadliest spot; even there as he aimed eagerly against him he smote him with the jagged stone, and he brake the bow-string; but his hand grew numb at the wrist, and he sank upon his knees and thus abode, and the bow fell from his hand; (*Iliad* Book 22, lines 324–326), but there was an opening where the collar bones part the neck and shoulders, even the gullet, where destruction of life cometh most speedily; even there, as he rushed upon him, goodly Achilles let drive with his spear; and clean out through the tender neck went the point.

Finally, Beasley briefly discusses the treatment of orthopedic injuries in both the Trojan and Greek camps. He ends by noting that when the Greek surgeon Machaon is wounded in the shoulder Greek hero Idomeneus calls to old man Nestor King of Pylos (*Iliad* Book 11, lines 511–516, translation by F. Horn, used with permission):

Nestor, son of Neleus, great glory of the Achaeans, come on, step on your chariot, and beside you shall Machaon step, and to the ships swiftly drive your single-hoofed horses: for a doctor is a man worth many others to cut out arrows and to sprinkle on soothing drugs.

Beasley could not agree more with this wisdom of Homer.

References

Altschuler, E. L. (2001). One of the oldest cases of schizophrenia in Gogol's Diary of a Madman. *BMJ (Clinical Research ed.), 323*(7327), 1475–1477.

Altschuler, E. L., Haroun, A., Ho, B., & Weimer, A. (2001). Did Samson have antisocial personality disorder? *Archives of General Psychiatry, 58*(2), 202–203.

Beasley, A. W. (1972). Homer and orthopaedics. *Clinical Orthopaedics and Related Research, 89*, 10–16.

Centers for Disease Control and Prevention (CDC). (2007). Nephrogenic fibrosing dermopathy associated with exposure to gadolinium-containing contrast agents – St. Louis, Missouri, 2002–2006. *MMWR. Morbidity and Mortality Weekly Report, 56* (7), 137–41.

Cowper, S. E., Robin, H. S., Steinberg, S. M., Su, L. D., Gupta, S., & LeBoit, P. E. (2000). Scleromyxoedema-like cutaneous diseases in renal-dialysis patients. *Lancet (London, England), 356*(9234), 1000–1001.

Cowper, S. E. (2003). Nephrogenic fibrosing dermopathy: The first 6 years. *Current Opinion in Rheumatology, 15*(6), 785–790.

Grobner, T. (2006). Gadolinium – A specific trigger for the development of nephrogenic fibrosing dermopathy and nephrogenic systemic fibrosis? *Nephrology, Dialysis, Transplantation, 21*(4), 1104–1108.

Homer, (1924). *The Iliad.* Harvard University Press.

Marckmann, P., Skov, L., Rossen, K., Dupont, A., Damholt, M. B., Heaf, J. G., & Thomsen, H. S. (2006). Nephrogenic systemic fibrosis: Suspected causative role of gadodiamide used for contrast-enhanced magnetic resonance imaging. *Journal of the American Society of Nephrology: JASN, 17*(9), 2359–2362.

Moschella, S. L., Kay, J., Mackool, B. T., & Liu, V. (2004). Case records of the Massachusetts General Hospital. Weekly clinicopathological exercises. Case 35-2004. A 68-year-old man with end-stage renal disease and thickening of the skin. *The New England Journal of Medicine, 351*(21), 2219–2227.

Thompson, R. C., Allam, A. H., Lombardi, G. P., Wann, L. S., Sutherland, M. L., Sutherland, J. D., Soliman, M. A.-T., Frohlich, B., Mininberg, D. T., Monge, J. M., Vallodolid, C. M., Cox, S. L., Abd el-Maksoud, G., Badr, I., Miyamoto, M. I., el-Halim Nur el-Din, A., Narula, J., Finch, C. E., & Thomas, G. S. (2013). Atherosclerosis across 4000 years of human history: The Horus study of four ancient populations. *Lancet (London, England), 381*(9873), 1211–1222. https://doi.org/10.1016/S0140-6736(13)60598-X.

Physicians and other medical personnel in Homer's epics

3

Christos Koutserimpas
School of Rehabilitation Health Sciences, University of Patras, Greece

Homer's epic poems, the *Iliad* and *Odyssey*, feature several notable figures who serve as healers and health providers. These heroes-characters play pivotal roles in Homer's narratives by attending to the wounds and ailments of the warriors and the main protagonists (Godquin, 1990; Koutserimpas et al., 2017). Trauma care and treatment cases are thoroughly described in Chapter 8.

In Homer's *Iliad*, the role of health providers is crucial in the intense environment of the battlefield, where they attend to the wounded and ensure the warriors can continue fighting Table 3.1. Mecisteus and Alastor are two notable figures among the Greeks who assist their comrades in moments of dire need. In Book 2, lines 721–725, Mecisteus and Alastor transport the wounded archer Teucer away from the battlefield, ensuring he is out of harm's way. Similarly, in Book 13, lines 421–423, they carry Hypsenor, who has also been injured, acting as modern-time paramedics (Koutserimpas & Samonis, 2017).

Machaon, another prominent healer, provides vital medical care to key Greek leaders. He is the chief surgical officer of the Greek camp and his importance is highlighted throughout *Iliad* (Koutserimpas & Samonis, 2018). His skills are highlighted in Book 4, verses 212–219, where he performs a surgical procedure on Menelaus, extracting an arrow from his abdomen and carefully treating the wound to prevent further damage. In Book 11, lines 510–515, Machaon attends to Nestor, providing both transportation and treatment for his injuries. These acts underscore Machaon's importance not only as a healer but also as a military officer in maintaining the morale and combat effectiveness of the Greek forces (Balanika & Baltas, 2014; Kakridis, 1986; Koutserimpas & Samonis, 2018).

On the Trojan side, Pelagon plays a similar role by assisting Sarpedon, who has been struck in the thigh with a spear. In Book 5, lines 692–698, Pelagon skillfully extracts the spear and provides necessary care to the wound, ensuring that Sarpedon can continue to lead and fight. Similarly, Glaucus, despite being injured himself, applies pressure to his own upper limb wound to control bleeding, as detailed in Book 16, lines 520–530. These actions highlight the vital importance of medical care in the heat of battle and the dedication of these individuals to their comrades, regardless of the immediate danger surrounding them (Koutserimpas et al., 2017).

In Homer's *Odyssey* health providers play a vital role in caring for characters and ensuring their well-being through various forms of medical and supportive care Table 3.2. Helen, the wife of Menelaus, demonstrates her knowledge of medicinal herbs in rhapsody δ, verses 220–227. She adds soothing herbs to the wine served to

Medicine in Homer. DOI: https://doi.org/10.1016/B978-0-443-32978-4.00003-1

Table 3.1 List of healthcare providers in the ***Iliad***. Individuals who provided healthcare during the Trojan War between the Greeks and the Trojans.

Health provider(s)	Patient(s)	Comments	Book/lines
Mecisteus and Alastor (Greeks)	1 Teucer (Greek) 2 Hypsenor (Greek)	Transport from the battlefield	1. 2/721–725 2. 13/421–423
Machaon (Greek)	1. Menelaus (Greek) 2. Nestor (Greek)	1. Surgical extraction of an arrow at the abdomen and wound care 2. Transport and injury care	1. 4/212–219 2. 11/510–515
Sthenelus (Greek)	Diomed (Greek)	Arrow extraction (shoulder injury)	5/112
Pelagon (Trojan)	Sarpedon (Trojan)	Spear extraction (thigh injury)	5/692–698
Diomed (Greek)	Diomed (Greek)	1. Wound care (shoulder injury) 2. Arrow extraction (foot injury)	1. 5/794–798 2. 11/396–397
Odysseus (Greek)	Odysseus (Greek)	Spear extraction (thorax injury)	11/455–458
Patroclus (Greek)	Eurypylus (Greek)	Arrow extraction and wound care (thigh injury)	11/844–849
Agenor (Trojan)	Helenus (Trojan)	Spear extraction and wound care (hand injury)	13/598–599
Hecamede (Greek)	Machaon (Greek)	Wound care and cleaning (shoulder injury)	14/1-3
Polydamas, Aeneas, Agenor, Sarpedon, Glaucus (Trojans)	Hector (Trojan)	Wound care and cleaning (thorax injury)	14/424–439
Glaucus (Trojan)	Glaucus (Trojan)	Pressure application to the trauma for bleeding control (upper limb injury)	16/520–530

Homer also mentions Gods with healing powers. A complete list of all wound care cases is provided in Chapter 8. Machaon stands out from the list as the chief military surgeon of the Greeks: (1) He attends to the serious wound of King Menelaus. (2) When Machaon himself is injured, all Greeks rally to him, as he was vital as a physician. Many other heroes, including Odysseus, Nestor and his slave-nurse Hecamede, possessed some medical knowledge. Author's own work.

their guests, including Telemachus and Nestor, alleviating their pain and stress. This act of hospitality and care underscores the importance of healing not just physical wounds but also emotional and psychological distress in the epic poem.

Eurycleia, the loyal nurse of Odysseus, holds a significant place in his life from birth. In rhapsody τ (Book 19), verses 354–355 and 482–483, she is described as helping during Odysseus's birth and assisting in his early care and lactation. Eurycleia's lifelong dedication to Odysseus reveals her role in his upbringing and later in recognizing him upon his return despite his disguised appearance.

Table 3.2 Healthcare providers in the Odyssey.

Health provider(s)	Patient(s)	Comments	Rhapsodies/ verses
Helen (Greek)	Guests of Menelaus including Telemachus, Nestor, and other Greeks	Herbs in the wine for pain	δ/220–227
Eurycleia (Greek)	Odysseus (Greek)	Midwife—helps in the birth and lactation of Odysseus	τ/354–355, 482–483
Autolykos (Odysseus' grandfather) and his children (Greeks)	Odysseus (Greek)	Wound care and bleeding stoppage	τ/453–460
Agamemnonas and warriors (Greeks)	Achilles (Greek)	Cleaning of the dead body with warm water and ointments	ω/41–46
The parents of mnesteres (Greeks) (narrative on what would happen if they were present)	Mnesteres (Greeks)	Cleaning of the dead bodies	ω/184–190

Author's own work.

Odysseus's grandfather, Autolykos, and his children also contribute to Odysseus's well-being, particularly in moments of injury. In Rhapsody τ (Book 19), verses 453–460, they provide wound care and stop the bleeding when Odysseus sustains an injury from a boar during a hunting expedition. Their care involves both practical medical treatment and ritualistic actions to ward off evil, highlighting the intersection of medical knowledge and cultural practices in healing. Similarly, in the final book, rhapsody ω, verses 41–46, Agamemnon along with other Greek warriors cleans Achilles' body with warm water and ointments, demonstrating respect and reverence for the dead.

The medical care and health practices depicted in both the *Iliad* and the *Odyssey* by Homer highlight the importance of healing and support in the ancient Greek world, both in times of war and peace. In the *Iliad*, the battlefield is a place of constant injury and death, where healers like Machaon, Mecisteus, and Alastor, among others, provide essential medical care, from surgical procedures to transport the wounded. Their actions are critical in maintaining the fighting capacity and morale of the warriors. These acts of medical care underline the importance of compassion and expertise in an environment dominated by violence and chaos.

References

Balanika, A. P., & Baltas, C. S. (2014). Hecamede: Homeric nurse of the battle-wounded in the Trojan War. *Journal of Medical Biography, 22*(1), 32–34. https://doi.org/10.1177/0967772013479261, http://www.jmb.rsmjournals.com.

Godquin, B. (1990). Was Homer a surgeon? *Chirurgie - Memoires de l'Academie de Chirurgie, 116*(2), 136–143.

Kakridis, I. (1986). *Greek mythology* (first edition). Ekdotike Athinon.

Koutserimpas, C., Alpantaki, K., & Samonis, G. (2017). Trauma management in Homer's Iliad. *International Wound Journal, 14*(4), 682–684. https://doi.org/10.1111/iwj.12672, http://onlinelibrary.wiley.com/journal/10.1111/(ISSN)1742-481X.

Koutserimpas, C., & Samonis, G. (2017). Letter to the editor: Medical structures during war in the Homeric era: Paramedics and 'expedition' hospitals. *Journal of Wound Care, 26*(3), 152–153. https://doi.org/10.12968/jowc.2017.26.3.152b.

Koutserimpas, C., & Samonis, G. (2018). Machaon: The first trauma surgeon in Western history? *Journal of Wound Care, 27*(10), 659–661. https://doi.org/10.12968/jowc.2018.27.10.659, http://www.internurse.com/cgi-bin/go.pl/library/issues.html?journal_uid=38.

Thersites' cleidocranial dysostosis and the story of the *Iliad* and *Odyssey*

4

Eric L. Altschuler[1,2]
[1]Essen Health Care, Bronx, NY, United States, [2]Kessler Institute for Rehabilitation, West Orange, NJ, United States

But, continuing from Chapter 2, there is one other character in the *Iliad* with a congenital orthopedic deformity that Beasley (Beasley, 1972) mentions. Unlike Hephaestus, Thersites is a mortal, not a god, and, furthermore, he is a commoner with no patronymic (father's name [from the Greek words πατήρ "father" and ὄνυμα "name"])[1] or place of origin listed by Homer—a *sui generis* case in the *Iliad*. Based on Thersites' ability to *approximate* his shoulders—bring his two shoulders together to the point of touching—due to an absence of collar bone and other deformities Beasley (Beasley, 1972) diagnosed Thersites with cleidocranial dysostosis (Cooper et al., 2001). In his survey of orthopedics in Homer Beasley seemed most interested and intrigued by Thersites—as was I when I read Beasley's paper. I was most excited to find a short note by a medical geneticist (Carter, 1973) published in the *Lancet* the year after Beasley's paper also diagnosed Thersites with cleidocranial dysostosis! Thersites' case got me interested in the whole topic of medicine in Homer (Altschuler, 2001).

Thirsites appears in only one episode in the *Iliad* (or the *Odyssey*) in Book 2, lines 211–277. To briefly summarize the action (Altschuler, 2001) until that point (see also Box 5.1 for a full synopsis of Book 1): the *Iliad* opens in the tenth year of the siege of Troy by the combined Greek forces under the command of Agamemnon, King of Mycenae. Agamemnon has captured Chryseis, daughter of Trojan priest Chryses. Chryses entreats Agamemnon to return his daughter, but to no avail. Then Chryses prays to the god Apollo (who is supporting the Trojans) who proceeds to shoot arrows into the Greek camp causing a plague of high mortality hitting first the mules and dogs, and then the humans. Agamemnon finally consents to return Chryseis. But because he does so, he demands that Achilles give him the woman Briseis whom he had captured. Achilles initially refuses to give up Briseis but finally does so. Then, Achilles leaves the Greeks to fight without him.

At this point shortly after the beginning of Book Two Thersites enters[2]:

[1] Even fathers of illegitimate sons (and daughter) are listed! See Fig. 18.2. Only Thersites is patronymic less.

[2] (Translation from Homer (1924). This work is licensed under a Creative Commons Attribution-ShareAlike 3.0 United States License.)

Medicine in Homer. DOI: https://doi.org/10.1016/B978-0-443-32978-4.00017-1

Now the others sate them down and were stayed in their places, only there still kept chattering on Thersites of measureless speech, whose mind was full of great store of disorderly words, wherewith to utter revilings against the kings, idly, and in no orderly wise, [215] but whatsoever he deemed would raise a laugh among the Argives. Evil-favored was he beyond all men that came to Ilios: he was bandy-legged and lame in the one foot, and his two shoulders were rounded, stooping together over his chest, and above them, his head was warped, and a scant stubble grew thereon. [220] Hateful was he to Achilles above all, and to Odysseus, for it was they twain that he was wont to revile; but now again with shrill cries he uttered abuse against goodly Agamemnon. With him were the Achaeans exceeding wroth, and had indignation in their hearts.

Howbeit with loud shoutings he spake and chid Agamemnon: [225] "Son of Atreus, with what art thou now again discontent, or what lack is thine? Filled are thy huts with bronze, and women full many are in thy huts, chosen spoils that we Achaeans give thee first of all, whensoever we take a citadel. Or dost thou still want gold also, [230] which some man of the horse-taming Trojans shall bring thee out of Ilios as a ransom for his son, whom I haply have bound and led away or some other of the Achaeans? Or is it some young girl for thee to know in love, whom thou wilt keep apart for thyself? Nay, it beseemeth not one that is their captain to bring to ill the sons of the Achaeans. [235] Soft fools! Base things of shame, ye women of Achaea, men no more, homeward let us go with our ships, and leave this fellow here in the land of Troy to digest his prizes, that so he may learn whether in us too there is aught of aid for him or no—for him that hath now done dishonor to Achilles, a man better far than he; [240] for he hath taken away, and keepeth his prize by his own arrogant act. Of a surety there is naught of wrath in the heart of Achilles; nay, he heedeth not at all; else, son of Atreus, wouldest thou now work insolence for the last time." So spake Thersites, railing at Agamemnon, shepherd of the host. But quickly to his side came goodly Odysseus, [245] and with an angry glance from beneath his brows, chid him with harsh words, saying: "Thersites of reckless speech, clear-voiced talker though thou art, refrain thee, and be not minded to strive singly against kings. For I deem that there is no viler mortal than thou amongst all those that with the sons of Atreus came beneath Ilios. [250] Wherefore 'twere well thou shouldst not take the name of kings in thy mouth as thou protest, to cast reproaches upon them, and to watch for home-going. In no wise do we know clearly as yet how these things are to be, whether it be for good or ill that we sons of the Achaeans shall return. Therefore dost thou now continually utter revilings against Atreus' son, Agamemnon, shepherd of the host, [255] for that the Danaan warriors give him gifts full many; whereas thou pratest on with railings. But I will speak out to thee, and this word shall verily be brought to pass: if I find thee again playing the fool, even as now thou dost, then may the head of Odysseus abide no more upon his shoulders, [260] nor may I any more be called the father of Telemachus, if I take thee not, and strip off thy raiment, thy cloak, and thy tunic that cover thy nakedness, and for thyself send thee wailing to the swift ships, beaten forth from the place of gathering with shameful blows."

[265] So spake Odysseus, and with his staff smote his back and shoulders; and Thersites cowered down, and a big tear fell from him, and a bloody weal rose up on

his back beneath the staff of gold. Then he sat him down, and fear came upon him, and stung by pain with helpless looks he wiped away the tear. [270] But the Achaeans, sore vexed at heart though they were, broke into a merry laugh at him, and thus would one speak with a glance at his neighbor: "Out upon it! verily hath Odysseus ere now wrought good deeds without number as leader in good counsel and setting battle in army, but now is this deed far the best that he hath wrought among the Argives, [275] seeing he hath made this scurrilous babbler to cease from his prating. Never again, I ween, will his proud spirit henceforth set him on to rail at kings with words of reviling."

Thersites seems to be the only character in the story who recognizes the absurdity of the fight between Agamemnon and Achilles and the doom and destruction it will likely cause the entire expedition.[3] Thersites' speech and the entire episode involving Thersites speaks to the main theme of the *Iliad*—Agamemnon's insistence on demanding Briseis from Achilles will severely and needlessly damage the Greek cause. But in lines 258 to 260 (see above) ". then may the head of Odysseus abide no more upon his shoulders, nor may I any more be called the father of Telemachus ..." also is expounded the main theme of the *Odyssey*—Odysseus' devotion to his son Telemachus.[4] (Interestingly, these lines (258–260) contain an ultra rare (Higbie, 1995) paedonymic (son's name [from the Greek words παῖς/παιδός "child" and ὄνυμα "name"]). The only other example in Homer in the *Iliad*, Book 4, line 354—again it is Odysseus referring to himself as the father of Telemachus.) The Thersites episode thus encompasses the main themes of both the *Iliad* and the *Odyssey*.

Cleidocranial dysostosis is a very rare (less than a thousand cases have been reported (Cooper et al., 2001)) autosomal dominantly inherited congenital skeletal disorder. Clinical features include clavicular aplasia and hypoplasia and the ability to approximate one's shoulders (pathognomonic for cleidocranial dysostosis when present), patent sutures and fontanelles, and supernumerary teeth (Cooper et al., 2001). The significance of clavicular defects in this disease was first discussed (other than of the *Iliad*) in 1765 (Martin, 1765) and there was probably not a complete accurate description of this disease until 1871 (Scheuthauer, 1871). The gene mutated in cleidocranial dysostosis was found to be the transcription factor CBFA1 (on chromosome 6p21) (Mundlos, 1999).[5] Consistent with the diagnosis of cleidocranial dysostosis are the prominent and important, although subtle and poetic, use of the shoulder motif in lines 258–260 and 265–266.

CCD is so rare one has to ask the question of how Homer could have known about the condition. It is so unusual that you cannot make it up. Homer must therefore have observed a case of CCD. Then the additional question must be asked about how

[3] Thersites, a commoner of Homer's invention speaking common sense, is perhaps in some ways "Homer's Falstaff."

[4] The *Odyssey* can be broken into three parts: The first four books are the so-called Telemarchy which discuss the education and coming to age of Odysseus' son Telemachus. Books 5–12 discuss the wanderings of Odysseus. The last 12 books (Books 13–24) cover Odysseus' return to Ithaca and his eventual reunion with Telemachus and his wife Penelope. (See also Box 21.1.)

[5] Some researchers studying CBFA1 may noy appreciate they are studying an "epic" gene!

someone with CCD not only got himself deployed with the Greek army but in such a noteworthy fashion that Homer wrote about him! As I wondered some years ago (Altschuler, 2001), did Homer himself have CCD[6]?

References

Altschuler, E. L. (2001). Cleidocranial dysostosis and the unity of the Homeric epics: an essay. *Clinical Orthopaedics and Related Research, 383*, 286–289.

Beasley, A. W. (1972). Homer and orthopaedics. *Clinical Orthopaedics and Related Research, 89*, 10–16.

Carter, C. O. (1973). Cleidocranial dysostosis in the Iliad. *Lancet (London, England), 2*(7824).

Cooper, S. C., Flaitz, C. M., Johnston, D. A., Lee, B., & Hecht, J. T. (2001). A natural history of cleidocranial dysplasia. *American Journal of Medical Genetics, 104*(1), 1–6.

Higbie, C. (1995). Heroes' names, Homeric identities. *Alfred Bates Lord Studies in Oral Tradition. 10.*

Martin, M. (1765). Sur un déplacement naturel de la clavicule. *Journal of Medicine of Clinical Pharmacology, 23*, 456–461.

Mundlos, S. (1999). Cleidocranial dysplasia: Clinical and molecular genetics. *Journal of Medical Genetics, 36*(3), 177–182.

Scheuthauer, G. (1871). Kombination rudimentärer Schlüsselbeine mit Anomalien des Schädels beim erwachsenen Menschen. *Allg Wien Med Ztg, 16*, 293–295.

[6] A third character possibly with a congenital orthopedic deformity seems to have escaped Beasley's notice: (*Iliad* Book 10, lines 315–317) Dolon son of Eumedes ill-favored to look upon, but withal swift of foot agreed to spy for the Trojans on the Greeks. Despite his speed, he is soon caught by wily Odysseus and Diomedes. Eventually, after giving up the plans of the Trojan war camp having been promised safety and rewards by Odysseus, he was decapitated by Diomedes (Table 13.1 injury 10). Is Dolon somehow Homer's Trojan counterpart of Thersites with a calling card of speed rather than perspicacity who winds up dead after being willing to trade his comrades' secrets for money?

Which translation of Homer should I use?

5

Eric L. Altschuler[1,2]

[1]Essen Health Care, Bronx, NY, United States, [2]Kessler Institute for Rehabilitation, West Orange, NJ, United States

If reading Homer in translation is good, hearing it in the original is always better. But understanding Greek is not so easy. Indeed, no less rhetorician than Cicero, as reported by Shakespeare in *Julius Cesar*, commented after attending a speech, "it was Greek to me." So is the situation therefore hopeless unless you already know Greek even to get a taste of Homer in the original? It is not! The key word here is Loeb, that is the Loeb Library parallel Greek/English translation of Homer.

Even in an era of e-books, reading on one's phone and bookstores where space for books must compete for space with espresso machines and pastry cases, the elegantly covered and bound green and red covered volumes for Greek and Latin classics, respectively, of the Loeb Classical Library persist not only in use but also on display Fig. 5.1.

AT Murray's definitive text and clean translation (Homer, 1924) of the *Iliad* is a classic and has been updated for modern readers by William Wyatt (Homer, 1999). (Similarly, George Dimock has updated Murray's translation of the *Odyssey*.) The volumes also include indications of lines that ancient scholars and grammarians (Zenadotus, Aristarchus) and writers (e.g., Plutarch) found corrupted (complete with proposed fixes/emendations/kluges) or rejected as well as the occasional note by Murray (updated as needed).

Even if you do not know any Greek, try reading a couple of pages a day in the parallel translation. You will get a sense of Homer's pace and style and you can dip your toe referencing the Greek by starting with proper names. As you learn more Greek (see below) you need Wyatt, Dimock, and Murray less!

Better than reading Homer in the parallel translation is reading Homer in Greek. But you ask how can you possibly read Homer in the original Greek if you cannot read Greek?! As above I have a one-word answer or in this case and one person answer, "Pharr."

If you took Latin in grammar school or high school, the name "Clyde Pharr" might ring a bell. Pharr is the guy who wrote the book with the iconic purple cover (Vergil's Aeneid: Books I-VI: With Introduction, Notes, Vocabulary, and Grammatical Appendix, 2024) and foldout list of the most common vocabulary words in Books 1–6 of Virgil's *Aeneid* (Pharr, 1930). On each page, the book also has a gloss of the less common words used in the text on that page text explanation, a complete Latin grammar in the back of the book, and also sections on rhetorical devices and the dactylic hexameter meter of Virgil's poetry. The book is indispensable to students

Medicine in Homer. DOI: https://doi.org/10.1016/B978-0-443-32978-4.00007-9

Figure 5.1 Loeb Classical Library: Greek volumes with green covers, Latin volumes with red covers. https://commons.wikimedia.org/wiki/Category:Loeb_Classical_Library#/media/File:Loeb_Classical_Library,_Waterstones,_Gower_Street.jpg By Philafrenzy—Own work, CC BY-SA 4.0, https://commons.wikimedia.org/w/index.php?curid=128106017.

and their teacher and is still in print nearly 95 years after its original publication! It turns out that Pharr wrote a book for students of Ancient Greek as well one that is particularly relevant for out purposes *Homeric Greek A Book for Beginners* (Pharr, 1920).

Students using Pharr's book on Virgil have already studied Latin for two or three years. His task in teaching Homer is much harder as students are complete beginning students who do not even know the Greek alphabet. But Pharr accomplishes the task! Starting with basic noun and verb paradigms in the front of the book and the alphabet in the back of the book in 77 efficient and entertaining lessons, Pharr takes the student through all the grammar of Homeric Greek, basic vocabulary, and then through all of Book 1 of the *Iliad*. (See Box 5.1 for description of the action in Book 1 (Homer's Iliad Summaries, 2024)).

Box 5.1 Synopsis of action in Book 1 of the Iliad

1–7 Prologue: "Muse, sing the rage of Achilles …."

8–32 The Trojan Priest of Apollo Chryses tries to ransom his daughter Chryseis who has been captured by the Greeks. Greek Commander-in-Chief Agamemnon roundly rejects Chryses' ransom offer and dismisses the priest.

33–52 Chryses prays to Apollo, who sends a plague among the Greek army.

53–67 After 10 days, Achilles summons an assembly.

68–100 Greek seer Calchas reveals the cause of Apollo's anger: Chryseis must be returned to her father.

101–147 Agamemnon reluctantly decides to give Chryseis back to her father but demands a replacement for Chryseis. He says in exchange he will take the girl Briseis who had been given to Achilles.

148–171 Achilles becomes angry and threatens to go back home.

172–187 Agamemnon decides he will take Achilles' prize, the girl Briseis.

188–222 Achilles is about to slay Agamemnon, but is stopped by the goddess Athena.

223–244 Achilles rages at Agamemnon. He declares that the Achaians will one day need his help but then he will not give it. He "takes his bat and ball and leaves the Greeks to play the game without him"—good luck—but does not leave the park.

245–284 Aged Nestor, King of Pylos, tries to make peace between Achilles and Agamemnon.

285–303 The entreaty is unsuccessful.

304–317 The assembly breaks up and Odysseus boards Chryseis on his ship to take her back to her father.

318–344 Agamemnon orders the heralds Talthybios and Eurybates to collect Briseis from Achilles.

345–363 Achilles cries to his mother the goddess ("calls her on his cellphone"). Thetis who asks why he is upset.

364–427 Achilles first says that his mom should already know why he is upset ("you should know why I'm upset"), and then proceeds to tell Thetis the reason for his outrage and despair. Achilles tells Thetis to ask Zeus to grant a great victory to the Trojans. Thetis notes that Zeus is away from Olympus for 12 days but promises her son that she will ask Zeus upon his return.

428–445 Odysseus brings Chryseis back to her father Chryses the Priest.

446–458 Chryses prays to Apollo to stop the plague, which he does.

459–487 Odysseus performs a sacrifice and returns home after a feast.

488–492 Achilles stays beside his ship.

493–530 Early in the morning the day of Zeus' return to Olympus Thetis meets with Zeus and Zeus then agrees to do what she asks despite knowing that his wife Hera will be furious.

531–567 Hera recognizes Thetis has been to see Zeus, is furious, and lets Zeus know it! Zeus doth protest way too much that Thetis had been to see him and angrily denies and disputes Hera's claim.

568–594 Hera's son Hephaistos, the lame blacksmith of the Gods, advises his mother to submit to Zeus, noting how he had been punished for interfering in one of their quarrels.

594–611 Hera submits to Zeus. Hephaistos plays the cupbearer during a party the Gods have on Olympus.

Pharr also includes maps, illustrations and pictures of sculptures and paintings. Though certainly not a linguistic books, interestingly, Pharr gives comments on origins and predecessors of Homer's Greek—the oldest Greek we have (except the Linear B tablets (Robinson, 2012))—and a dialect that scholars can still not place temporally, geographically or say if it ever was even spoken. Pharr also explains and gives examples of how to SCAN the meter (DACTYLIC HEXAMETER) of Homer's *Iliad* and *Odyssey* (see Box 5.2). (David Chamberlain has recorded the entire *Iliad* and almost ⅓ of the *Odyssey* and his excellent website (Greek and Roman Verse, 2017) which lets you play back the poem line by line.)

Like his book on Virgil, Pharr's *Homeric Greek for Beginners* is also still in print though in this case in a revised edition differing from the original mostly by cutting out Biblical references in the original. I do not find these particularly pious, sectarian or proselytizing, but rather touching on subjects such as prayers and sacrifices to Gods and interactions and conflicts among peoples that are found in the *Iliad* and other ancient and modern societies. Many of Pharr's references are from the so-called "J" or Yahwist source of the Bible (compared to Homer in (Auerbach, 1953)) or extended J text (Altschuler et al., 2001a; Altschuler et al., 2001b; Friedman, 1998). You can find and use the original edition on Google Books (Pharr, 1920). It is even more fun, exciting, inspirational, and educational to go to the internet, buy a hard copy which typically has the prior learner's study points (and minor corrections to the book), and learn from the prior owner's notes and to spot a correction here or there by yourself!

In junior high school, "Pharr" was really more of an adjective—the Pharr book—than a proper noun. If we ever thought of Pharr as a person we probably imaged him as a great teacher, like our teacher Mrs. Ellis, who happened to write a good textbook with the foldout list of common vocabulary words and the always welcome vocabulary glosses at the bottom of each page for the uncommon words. When I started using Pharr's Homer book in high school to learn Homeric Greek I noticed that it was published by a university press so I assumed then that Pharr must be a Professor somewhere, not that I really knew much about what professors did. Coming back to Pharr much later I was intrigued, impressed at somewhat surprised to read the discussion Pharr has on the Greek language before Homer including points that I did not recall learning in my Greek or Hittite classes in college. Also, his sparing, but on point reference of proto-Indo-European forms clearly marked him to me as a scholar as well as a master educator. Furthermore, Pharr's Biblical references seemed not at all religious but rather apt parallels reflecting a keen insight into Homer's worldview and thinking. I was thrilled to have these opinions of mine about

Box 5.2 Homer's meter

The language of Homer, like that of Shakespeare (predominantly) is poetry. Both write in metered free verse—rhyming not obligatory but only used rarely for effect—but with different meters. Homer's two epics are written in DACTYLIC HEXAMETER Shakespeare's in IAMBIC PENTAMETER. Hexameter (pentamer) means that each line has six (hex-) (five (pent-)) feet (meter). A dactyl is a foot consisting of a long syllable followed by two short syllables: – ◡◡. (An iamb consists of a short syllable followed by a long syllable: ◡ –).

In Homer's dactylic hexameter the first four feet are dactyls or spondees—two long syllables (– –) thus lasting the same length as a dactyl. The large majority of the time the fifth foot in Homer is a dactyl and the sixth foot can be a spondee or a trochee (a long syllable followed by a short syllable (– ◡). (The last syllable can be scanned agnostically as to long or short and marked "x.") Lines in which the fifth foot is a spondee (SPONDAIC LINES) are reserved by Homer to emphasize solemn situations or points. In brief, syllables in the epics are long if they contain a long vowel such as an eta (H,η; pronounced "ay" as in English awAY), or a vowel followed by two consonants. A full discussion of the assignment of syllable lengths in Homer is beyond the scope of this book but is given in Pharr (1920, sections 1159–1192) and in other texts on Homer.

Dactylic hexameter is very rarely used for English verse but the example below by Longfellow is illustrative of this meter (Fig. 5.2).

Next is an example from Homer (see Fig. 5.3) with the long and short syllables marked of the feet numbered. Notice that line 40 is a spondaic line.[3]

Translation:

For he [Agamemnon] believed that he would take the city of Priam on this day, the fool, and he did not know what deeds Zeus was planning for he was yet to bring pains and groanings on both the Trojans and the Danaans in mighty battles. And he woke from his sleep, and the divine voice flowed around him.

(Translation by Fabian Horn.)

Along with the twenty-four letters of the Greek alphabet shown in Table 1.1, the Homeric Epics contain more than a ghostly trace of a twenty-fifth letter—digamma (F, ϝ)—that can half a significant impact on the meter of a line. Pronounced like an English "w," the digamma dates from an era of Greek somewhat before the time Homer was writing but still makes an impact in the

[3] For comparison, given below are lines 81–85 from Act 1, Scene 4 of Romeo & Juliet (1597 Quarto) which describe King Mab's nightmare awakenings due to PTSD. In English poetry, unlike Greek poetry, syllables are denoted "long" or "short" not based on vowel length but on being accented (long) or unaccented (short). It is left as an exercise for the reader to scan Shakespeare's iambic pentameter: And then dreams he of cutting foreign throats,/ Of breaches, ambuscados, countermines,/ Of healths fiue fadome deep, and then anon/ Drums in his eare: at which he startes and wakes,/ And swears a Praier or two and sleepes againe.

— ∪ ∪ — ∪ ∪ — ∪ ∪ — ∪ ∪ — ∪ ∪ — —

This is the | forest prim | eval. The | murmuring | pines and the | hemlocks,

— ∪ ∪ — ∪ ∪ — — — ∪ ∪ — ∪ ∪ — —

Bearded with | moss, and in | garments | green, indis | tinct in the | twilight,

— — — ∪ ∪ — — — — — ∪ ∪ — —

Stand like | Druids of | eld, with | voices | sad and pro | phetic,

Figure 5.2 Longfellow scanned. Longfellow's Evangeline, A Tale of Acadie (1847), lines 1–3 with dactylic hexameter scanned. Text Public Domain, meter scansion by E. Altschuler.

Iliad 2,37-41:

1— ∪ ∪ | 2— —| 3— ∪∪|4— ∪ ∪|5— ∪ ∪ |6— x
φῆ γὰρ ὅ γ' αἱρήσειν Πριάμου πόλιν ἤματι κείνῳ

1— ∪∪ | 2— ∪ ∪|3— ∪ ∪|4— — | 5— ∪ ∪|6— x
νήπιος, οὐδὲ τὰ ᾔδη ἅ ῥα Ζεὺς μήδετο ἔργα·

1— — | 2— ∪ ∪|3— ∪ ∪|4— ∪∪|5— ∪ ∪|6— x
θήσειν γὰρ ἔτ' ἔμελλεν ἐπ' ἄλγεά τε στοναχάς τε

1— ∪ ∪ | 2— ∪ ∪|3—∪ ∪|4— ∪ ∪|5— —|6— x
Τρωσί τε καὶ Δαναοῖσι διὰ κρατερὰς ὑσμίνας.

1— ∪ ∪ | 2— — | 3— —|4— ∪ ∪|5— ∪ ∪ |6— x
ἔγρετο δ' ἐξ ὕπνου, θείη δέ μιν ἀμφέχυτ' ὀμφή·

Figure 5.3 Iliad Book 2, lines 37–41 with meter scanned. Iliad Book 2, lines 37–41 with the long and short syllables and the feet of the dactylic hexameter marked. *Source*: Meter scanning by F. Horn. Used with permission.

Iliad and the *Odyssey*. For example the Homeric Greek work ἔργον ("work," "deed"; plural ἔργα) was originally ϝἔργον. In line 38 from the example above the digamma has no mere shadowy effect on the poetry: to eye (in print) there appears to be what is known as a HIATUS occurring between the last two words. That is, the second to last word of the line ends in a vowel (omicron) and the last word ἔργα begins with a vowel (epsilon). Such a situation does not flow well poetically and is typically remedied by ELISION (omitting the final vowel of the first word of the potential hiatus) or adding a so-called MOVEABLE CONSONANT such as a ν (nu) or ς (sigma) at the end of the first word of the hiatus. But here the invisible presence of the digamma is felt and neither of these is necessary!

Unlike Shakespeare who on occasion uses prose (and very rarely had a "long line" hexamer) the *Iliad* and the *Odyssey* are exclusively in dactylic hexameter. Later (see Box 21.2) we will see how this restriction of the meter of the epics may shed some light on Homer's compositional process.

Pharr totally confirmed and validated when I came across a paper "Into Thin Air: Shakespeare's The Tempest and Virgil" published in the journal *Notes and Queries* by Robert Easting of Victoria University in Wellington (Easting, 2023).

While the paper was published in the June issue of *Notes and Queries*, I actually read the paper in May. How you may ask did I read a paper published in November in May? N&Q is one of the journals I check the Advance online articles every week to stay abreast of developments in the humanities. (N&Q established in London in 1849 is one of the oldest English literature and language [broadly understood!] journals still in publication older even than *Nature* (1869), though not quite as old as *The New England Journal of Medicine* (1812 originally *New England Journal of Medicine and Surgery and the Collateral Branches of Medical Science*) and *The Lancet* (1823).) Somehow, I find it less depressing when reading a paper with an important discovery that I did not discover it to read it Advance online.

My Jedi sense told me to read this whole paper and not just the title. (Granted, the journal also made it available free full text so it only took a double click to read the entire paper.)

Easting is looking at the origin of the phrase "into thin air." He notes that according to the *Oxford English Dictionary* (OED), the first use is by Shakespeare in *The Tempest* (Act 4, Scene 1, lines 148–150) "These our actors. were all Spirits, and / Are melted into Ayre, into thin Ayre."[1] Easting thinks that Shakespeare got the phrase from Virgil's *Aeneid* Book IV, Lines 276–278:

tali Cyllenius ore locutus
mortalis visus medio sermone reliquit
et procul **in tenuem** ex oculis evanuit **auram**

With these words on his lips, Mercury vanished Into thin air, visible no more to human eyes

(English translation from (Virgil, 2005)).

Easting makes various philological points to bolster his case that Shakespeare got the phrase from Virgil. Additionally, and very convincing are the multiple connections between the Tempest and Virgil that Easting points out surely making very likely that SS's source was Virgil.

Easting concludes by saying that neither the OED or Shakespeare scholars have recognized that the honor of priority for use of the phrase "into thin air" should go to Virgil.

But hold on. In a footnote after the final period of the paper (Easting, 2023) Easting notes that "one editor of Virgil simply annotate *Aeneid* IV. 278 'cf. Shakespeare, *Tempest*, IV, I, 148–150.'" Who is that editor? Our man Pharr in his Vergil textbook (Pharr, 1930)[2]! Pharr has still got it and he always had it as a scholar and a teacher!

[1] Easting also notes "vanish into air," from Othello (Scene 3, Act 1, line 20).

[2] Pharr compares Achilles' comment in line 167 of Book 1 of the Iliad that his prize the girl Briseis whom Agamemnon is taking away from him (see Box 5.1) is "of slight value yet dear (to me) [ὀλίγον τε φίλον τε]" to a line from Shakespeare's As You Like I (Act 5, scene 4) "A poor virgin, Sir,/an ill-favored thing, Sir, but mine own."

So, start with your favorite translation. Eventually, it will be your time to read and learn Homeric Greek from Pharr's book. Once you've finished this and gone through Book 1 with Pharr as your guide and teacher, get the Loeb parallel translation and move up to reading the *Iliad* at least partly in Greek. Also soon then start learning Homeric vocabulary using Pharr's (of course!) user friend version of Owen and Goodspeed's vocabulary lists (Owen & Goodspeed, 1969) and begin reading more of the Greek and less of the parallel English. If you need to look up an uncommon word Richard Cunliffe's A Lexicon of the Homeric Dialect (Cunliffe, 1924)is an invaluable resource. Eventually start to use the app on the companion website and hear from Homer himself!

If you need help, for any book in the *Iliad* or *Odyssey*, Pharr's door is open for office hours!

References

Altschuler, E. L., Haroun, A., Ho, B., & Weimer, A. (2001a). Did Samson have antisocial personality disorder? *Archives of General Psychiatry, 58*(2), 202–203.

Altschuler, E. L., Haroun, A., Ho, B., & Weimer, A. (2001b). Did Samson have antisocial personality disorder? *Archives of General Psychiatry, 59*(6), 565–566.

Auerbach, E. (1953). *Mimesis: The representation of reality in western literature*. Princeton University Press.

Cunliffe, R. J. (1924). *A Lexicon of the Homeric dialect*. Blackie and Son Limited.

Easting, R. (2023). Into thin air: Shakespeare's The Tempest and Virgil. *Notes and Queries, 70*(2), 79–80. https://doi.org/10.1093/notesj/gjad033.

Friedman, R. E. (1998). *The hidden book in the Bible*. HarperSanFrancisco.

(2017). Greek and Roman Verse https://hypotactic.com/my-reading-of-homer-work-in-progress/.

Homer (1924). The Iliad: Harvard University Press.

Homer (1999). The Iliad: Harvard University Press.

(2024). Homer's Iliad summaries https://www.geneseo.edu/~walker/Iliad%20Summary.htm.

Owen, W. B., & Goodspeed, E. J. (1969). *Homeric vocabularies*. Oklahoma University Press.

Pharr, C. (1920). *Homeric Greek: A book for beginners*. D. C. Heath Co.

Pharr, C. (1930). *Vergil's Aeneid: Books I-VI: With introduction, notes, vocabulary, and grammatical appendix*. D. C. Heath and Company.

Robinson, A. (2012). *The man who deciphered linear B: The story of Michael Ventris*. Thames & Hudson.

Vergil's Aeneid: Books I-VI: (2024). With introduction, notes, vocabulary, and grammatical appendix, https://www.bolchazy.com/Vergils-Aeneid-Books-I-VI-With-Introduction-Notes-Vocabulary-and-Grammatical-Appendix-P3690.aspx.

Virgil (2005). *Virgil: Aeneid*. Hackett Publishing Company.

Section 2

Trauma in homer

A trauma registry of Homer's *Iliad* 6

Christos Koutserimpas
School of Rehabilitation Health Sciences, University of Patras, Greece

Introduction

The Homeric Epics, the *Iliad* and the *Odyssey*, stand as the pinnacle of intellectual achievement within early ancient Greek civilization, earning their place among the masterpieces of world literature (Homer, n.d.; *Kakridis I: Greek mythology*, 1986).

Trauma is primarily depicted in the *Iliad* which recounts the saga of the Trojan War involving a ten-year siege and the battles between the Achaeans (Greeks) and Trojans at Troy's walls. Troy, also known as Ilion, lends its name to the Epic. According to Greek mythology, Helen, the beautiful wife of King Menelaus of Sparta, was abducted by Paris, the son of King Priam of Troy. To reconcile the offense, a coalition of powerful ancient Greek kings, led by King Agamemnon of Argos, brother of King Menelaus, organized a military campaign against Ilion (Homer, n.d.).

The initial documentation of trauma during the Trojan War was performed by H. Frölich, a German military doctor, in 1879 (Fröhlich, 1879). A detailed account of 147 injuries was provided, categorized as follows: 31 head injuries (21%), 16 neck injuries (11%), 79 trunk injuries (54%), 10 upper extremity wounds (7%), and 11 lower extremity wounds (7%) (Fröhlich, 1879). The majority of these injuries proved to be fatal (77.5%). The head injuries recorded the highest mortality rate at 100%, followed by trunk injuries at 85% and neck injuries at 81%. Conversely, extremity wounds displayed a significantly lower mortality rate at 14%. The primary causes of wounds were spear-related (106), followed by sword (17) and arrow (12) injuries. Additionally, 12 wounds from slingshots were documented. Among these weapons, the sword and spear exhibited the highest lethality, with mortality rates of 100% and 80% respectively, followed by slingshot and arrow injuries at 67% and 42% mortality rates, respectively (Fröhlich, 1879).

The purpose of the present chapter is to present a detailed trauma registry of the *Iliad* by Homer, by thoroughly evaluating the poem's text in its original form from a medical standpoint.

Methods

The Greek text of the *Iliad* by Homer was meticulously examined in its original to identify each instance of trauma. Every Rhapsody (Book) and verse were scrutinized to determine details such as the type and location of the injury, the assailant, the

Medicine in Homer. DOI: https://doi.org/10.1016/B978-0-443-32978-4.00016-X

victim, the weapon employed, and the eventual outcome of the wound. Cases that were repeated were included once in the listing.

Results—trauma registry

Table 6.1 provides the complete trauma registry of Homer's *Iliad*.

Table 6.2 provides the listing of all injuries that were mentioned in the *Iliad* which did not occur at the Trojan War but refer to past events.

A total of 249 trauma cases were recorded (234 injuries that occurred during the siege of Troy, Table 6.1, and 15 from past events, Table 6.2), including more than 273 lethal injuries (some cases refer to multiple deaths). Regarding the anatomical region of these wounds, the majority of the wounds (65; 26.1%) were located at the trunk (back, thorax, and abdomen), while 59 of them (23.7%) were at the head and neck, 29 (11.6%) at the upper extremity (scapula, shoulder, arm, forearm, wrist, and hand) and 14 (5.6%) at the lower extremity (hip, groin area, thigh, and knee). The heroes described as the "attacker" in most of the cases were the Greek heroes; Achilles (in 26 instances; 10.4%), Diomedes (21; 8.4%); and Ajax (19; 7.6%).

A total of 178 (71.5%) sharp weapon injuries (SWIs) and 24 (9.6%) blunt trauma cases were identified. The spear was the most commonly used sharp weapon, used in 116 instances (65.2%), followed by the sword in 29 (16.3%), while arrows were involved in 19 cases (10.7%).

The majority of these traumas were lethal (182 cases; 73.1%). SWIs were fatal in most cases (136 out of 178 instances; 76.4%), while 6 (25%) blunt traumas led to death. The trunk injuries revealed an 84.6% mortality rate, while the head and neck traumas were 86.4%. Moreover, the upper and lower extremity injuries had 20.7% and 35.7% mortality rates, respectively.

Discussion

The Homeric poems, the *Iliad* and the *Odyssey*, represent the zenith of intellectual prowess in early ancient Greek society, solidifying their status among the greatest works in global literature (*Kakridis I: Greek mythology*, 1986; Passas, 1960). Specifically, the *Iliad* captures pivotal moments from the siege's final weeks, emphasizing the intense conflict between King Agamemnon and the famed warrior, Achilles.

In the late 19th century, Heinrich Schliemann, an amateur archeologist, unearthed the remains of Troy in Asia Minor, providing a potential historical basis for the epic poem. The *Iliad* consists of 24 "rhapsodies" or "books," each designated by a letter of the Greek alphabet.

In the literature, various counts of total injuries have been documented. Fröhlich seems to be the first to comprehensively outline the wounds within The *Iliad* in 1879 (Fröhlich, 1879). Fröhlich identified 147 total wounds in the *Iliad* (Fröhlich, 1879). Divergent figures were presented by Santos (2000) (130 wounds), Apostolakis et al. (2010) (151 wounds), and other investigators, some elevating the count to 184

Table 6.1 The complete trauma registry of the injuries during the Trojan War, from the *Iliad*.

Case number	Rhapsody/ verses	Anatomical region	Type of trauma	Weapon	Attacker	Defender (injured)	Outcome
1.	A/50–51	–	SWI	Arrows	Apollo (God)	Greek warriors	Death
2.	A/382–383	–	SWI	Arrows	Apollo (God)	Greek warriors	Death
3.	A/591–594	–	Blunt trauma	Fall	Zeus (God)	Hephestus (God)	Survival
4.	B/265–269	Shoulder and scapula	Blunt trauma	Scepter	Odysseus (Greek)	Thersitis (Greek)	Survival
5.	B/701	–	SWI (verb κτείνω (slay used to describe injury)	–	Dardanos (Trojan)	Protesilas (Greek)	Death
6.	B/721–724	Trunk (Body)	Bite	–	Snake	Philoctetes (Greek)	Survival
7.	B/860–861	–	–	–	Achilles (Greek)	Ennomos (Trojan)	Death
8.	B/874–875	–	–	–	Achilles (Greek)	Nastis (Trojan)	Death
9.	Γ/370–375	Throat	Blunt trauma	Helmet's strap	Menelaus (Greek)	Paris (Trojan)	Survival
10.	Δ/134–140	Thorax	SWI	Arrow	Pandaros (Trojan)	Menelaus (Greek)	Survival
11.	Δ/457–462	Head	SWI ("αἰχμὴ χαλκείη" bronze point)	–	Antilochus (Greek)	Thalasiadis Echelopos (Trojan)	Death
12.	Δ/466–469	Thorax (ribs)	SWI	Spear	Aginor (Trojan)	Elephinor (Greek)	Death
13.	Δ/473–481	Thorax and shoulder	SWI	Spear	Ajax (Greek)	Simoisios (Trojan)	Death

(*Continued*)

Table 6.1 (Continued)

Case number	Rhapsody/ verses	Anatomical region	Type of trauma	Weapon	Attacker	Defender (injured)	Outcome
14.	Δ/489–493	Groin	SWI	Spear	Antiphos (Trojan)	Leukos (Greek)	Death
15.	Δ/496–503	Head (Face)	SWI	Spear	Odysseus (Greek)	Dimokontas (Trojan)	Death
16.	Δ/517–524	Lower extremity	Blunt trauma	Rock	Piros (Trojan)	Doris (Greek)	Survival
17.	Δ/524–526	Abdomen	SWI	Spear	Piros (Trojan)	Doris (Greek)	Death
18.	Δ/527–529	Thorax	SWI	Spear	Thoas (Greek)	Piros (Trojan)	Survival
19.	Δ/530–531	Abdomen	SWI	Sword	Thoas (Greek)	Piros (Trojan)	Death
20.	E/17–19	Thorax	SWI	Spear	Diomedes (Greek)	Figeas (Trojan)	Death
21.	E/40–42	Back and Thorax	SWI	Spear	Agamemnonas (Greek)	Odios (Trojan)	Death
22.	E/45–47	Shoulder	SWI	Spear	Idomenias (Greek)	Festos (Trojan)	Death
23.	E/49–58	Back and Thorax	SWI	Spear	Menelaus (Greek)	Skamandrios (Trojan)	Death
24.	E/65–68	Gluteus and abdomen	SWI	Spear	Myrionis (Greek)	Fereklos (Trojan)	Death
25.	E/69–75	Head and Neck	SWI	Spear	Megis (Greek)	Pidasos (Trojan)	Death
26.	E/76–83	Shoulder	SWI	Sword	Eyripilos (Greek)	Ipsinoras (Trojan)	Death
27.	E/95–100	Shoulder	SWI	Arrow	Pandaros (Trojan)	Diomedes (Greek)	Survival
28.	E/144	Thorax	SWI	Spear	Diomedes (Greek)	Astinos (Trojan)	Death

29.	E/144–147	Shoulder and neck	SWI	Sword	Diomedes (Greek)	Hyperonas (Trojan)	Death
30.	E/147–151	–	–	–	Diomedes (Greek)	Avantas and Polidos (Trojans)	Deaths (x2)
31.	E/152–155	–	–	–	Diomedes (Greek)	Xanthos and Thonas (Trojans)	Deaths (x2)
32.	E/160–164	Neck (location of injury inferred from Homer's simile: Even as a lion leapeth among the kine and breaketh the neck of a heifer or a cow as they graze in a woodland pasture, so did Tydeus' son [Diomedes] thrust both these in evil wise from their car, sorely against their will, and thereafter despoiled)	–	–	Diomedes (Greek)	Chromios and Echemonas (Trojans)	Deaths (x2)
33.	E/290–294	Head	SWI	Spear	Diomedes (Greek)	Pandaros (Trojan)	Death
34.	E/302–310	Abdomen and Hip	Blunt trauma	Rock	Diomedes (Greek)	Aeneas (Trojan)	Survival
35.	E/335–340	Upper extremity (Hand)	SWI	Spear	Diomedes (Greek)	Aphrodite (Goddess)	Survival
36.	E/385–390	Upper extremity	Blunt trauma	Rope	Efialtis and Otos	Ares (God)	Survival

(*Continued*)

Table 6.1 (Continued)

Case number	Rhapsody/ verses	Anatomical region	Type of trauma	Weapon	Attacker	Defender (injured)	Outcome
37.	E/392–394	Thorax	SWI	Arrow	Son of Amphitrionas (Greek)	Hera (Goddess)	Survival
38.	E/395–402	Shoulder	SWI	Arrow	Son of Zeus (Hemigod)	Hades (God)	Survival
39.	E/533–534	Abdomen	SWI	Spear	Agamemnomas (Greek)	Diikontis (Trojan)	Death
40.	E/541–542	–	SWI	Sword	Aeneas (Trojan)	Krithonas and Orsilochus (Greeks)	Deaths (x2)
41.	E/576–579	Neck	SWI	Spear	Menelaus (Greek)	Polimenis (Trojan)	Death
42.	E/582–588	Elbow	Blunt trauma	Rock	Antilochus (Greek)	Midonas (Trojan)	Survival
43.	E/582–588	Head	SWI	Sword	Antilochus (Greek)	Midonas (Trojan)	Death
44.	E/608–609	–	–	–	Hector (Trojan)	Menesthos and Agchialus (Greeks)	Deaths (x2)
45.	E/610–616	Abdomen	SWI	Spear	Ajax (Greek)	Amphios (Trojan)	Death
46.	E/656–663	Neck	SWI	Spear	Sarpidon (Trojan)	Tlipolemos (Greek)	Death
47.	E/656–663	Thigh	SWI	Spear	Tlipolemos (Greek)	Sarpidon (Trojan)	Survival

48.	E/677	–	SWI ("χαλκῷ"—"with bronze")	–	Odysseus (Greek)	Kiranos, Alastor, Alkandros, Noimonas, Pritanis, Alios, Chromeos (Trojans)	Deaths (x7)
49.	E/704–707	–	–	–	Hector (Trojan) and Ares (God)	Teuthra, Orestis, Trichos, Oenomaus, Aetolios, Elenos, Oresvis (Greeks)	Deaths (x7)
50.	E/844–848	–	SWI (verb κτείνω (slay used to describe injury)	–	Ares (God)	Periphas (Greek)	Death
51.	E/858–871	Groin	SWI	Spear	Diomedes (Greek)	Ares (God)	Survival
52.	Z/7–11	Head	SWI	Spear	Ajax (Greek)	Acamantas (Trojan)	Death
53.	Z/12–13	–	–	–	Diomedes (Greek)	Axilos (Trojan)	Death
54.	Z/17–19	–	–	–	Diomedes (Greek)	Calisis (Trojan)	Death
55.	Z/20–21	–	–	–	Eyryalos (Greek)	Opheltios and Drisos (Trojans)	Deaths (x2)
56.	Z/27–28	–	–	–	Mikistiadis (Greek)	Pidasos and Aesipos (Trojans)	Deaths (x2)

(Continued)

Table 6.1 (Continued)

Case number	Rhapsody/ verses	Anatomical region	Type of trauma	Weapon	Attacker	Defender (injured)	Outcome
57.	Z/29	–	–	–	Polypitis (Greek)	Astyalos (Trojan)	Death
58.	Z/30	–	–	–	Odysseus (Greek)	Perkosiotis (Trojan)	Death
59.	Z/31	–	SWI	Spear	Teukros (Greek)	Aretaos (Trojan)	Death
60.	Z/32	–	SWI	Spear	Antilochus (Greek)	Avliros (Trojan)	Death
61.	Z/33	–	–	–	Nestoridis and Agamemnom-as (Greeks)	Elatos (Trojan)	Death
62.	Z/35	–	SWI	Sword	Litos (Trojan)	Philakos (Greek)	Death
63.	Z/36	–	–	–	Evrypylos (Greek)	Melanthis (Trojan)	Death
64.	Z/63–64	Groin	SWI	Spear	Agamemnonas (Greek)	Adrastos (Trojan)	Death
65.	H/8–9	–	–	–	Paris (Trojan)	Menesthius (Greek)	Death
66.	H/11–12	Neck	SWI	Spear	Hector (Trojan)	Ionias (Greek)	Death
67.	H/15–16	Shoulder	SWI	Spear	Glaucos (Trojan)	Ifinos (Greek)	Death
68.	H/261–263	Neck	SWI	Spear	Ajax (Greek)	Hector (Trojan)	Survival
69.	Θ/119–123	Thorax	SWI	Spear	Diomedes (Greek)	Son of Iniopeas (Trojan)	Death
70.	Θ/256–260	Thorax and Back	SWI	Spear	Diomedes (Greek)	Achelaus (Trojan)	Death

71.	Θ/273–279	–	SWI	Arrow	Teukros (Greek)	Lycofontis, Melanippos, Amopaonas, Orsilochus, Ormenos, Ofelesis and Chromios (Trojans)	Death (x7)
72.	Θ/302–308	Thorax	SWI	Arrow	Teukros (Greek)	Gorgythos (Trojan)	Death
73.	Θ/309–315	Thorax	SWI	Arrow	Teukros (Greek)	Archipolemos (Trojan)	Death
74.	Θ/320–329	Shoulder (paralyzed upper extremity due to nerve injury)	Blunt trauma	Rock	Hector (Trojan)	Teukros (Greek)	Survival
75.	K/454–456	Neck	SWI	Sword	Diomedes (Greek)	Dolonas (Trojan)	Death
76.	K/482–485	–	SWI	Sword	Diomedes (Greek)	12 men from Thrace	Death (x12)
77.	K/494–497	–	SWI	Sword	Diomedes (Greek)	Son of Onidis (Trojan)	Death
78.	Λ/91–93	–	SWI	Spear	Agamemnonas (Greek)	Vienoras (Trojan)	Death
79.	Λ/93–98	Head	SWI	Spear	Agamemnonas (Greek)	Horse rider (Trojan)	Death
80.	Λ/107–109	Thorax	SWI	Spear	Agamemnonas (Greek)	Isos (Trojan)	Death
81.	Λ/107–109	Head	SWI	Sword	Agamemnonas (Greek)	Antifos (Trojan)	Death

(Continued)

Table 6.1 (Continued)

Case number	Rhapsody/verses	Anatomical region	Type of trauma	Weapon	Attacker	Defender (injured)	Outcome
82.	Λ/143–145	Thorax	SWI	Spear	Agamemnonas (Greek)	Pisandros (Trojan)	Death
83.	Λ/145–148	Throat	SWI	Sword	Agamemnonas (Greek)	Ippolochus (Trojan)	Death
84.	Λ/158–159	Head and neck	SWI	–	Agamemnonas (Greek)	Multiple Trojan warriors	Death (several)
85.	Λ/238–240	Neck	SWI	Sword	Agamemnonas (Greek)	Iphidamas (Trojan)	Death
86.	Λ/251–254	Forearm	SWI	Spear	Kononas (Trojan)	Agamemnonas (Greek)	Survival
87.	Λ/260–261	Head and neck (Decapitation)	SWI	Spear and Sword	Agamemnonas (Greek)	Kononas (Trojan)	Death
88.	Λ/302–309	–	SWI	–	Hector (Trojan)	Ipponoos, Aesymnos, Oros, Agelaos, Ophelios, Dolopas, Klytidis, Opitis, Antonoos, Asaeos (Greeks)	Death (x10)
89.	Λ/320–321	Thorax	SWI	Spear	Diomedes (Greek)	Thymvraios (Trojan)	Death

90.	Λ/321–322	–	SWI	Spear	Odysseus (Greek)	Molionas (Trojan)	–
91.	Λ/333–334	–	SWI	Spear	Diomedes (Greek)	Ippodamas (Trojan)	Death
92.	Λ/335	–	SWI (ἐξενάριξεν (slay))	–	Odysseus (Greek)	Ypeirochos (Trojan)	Death
93.	Λ/338	Hip	SWI	Spear	Diomedes (Greek)	Agastrophos (Trojan)	Death
94.	Λ/354–360	Head (Helmet)	Blunt trauma	Spear	Diomedes (Greek)	Hector (Trojan)	Survival
95.	Λ/376–378	Lower extremity	SWI	Arrow	Paris (Trojan)	Diomedes (Greek)	Survival
96.	Λ/420–421	Shoulder	SWI	Spear	Odysseus (Greek)	Diopitis (Trojan)	–
97.	Λ/422	–	–	–	Odysseus (Greek)	Ennomos, Thoas (Trojans)	Deaths (x2)
98.	Λ/423	Abdomen	SWI	Spear	Odysseus (Greek)	Hersidamas (Trojan)	Death
99.	Λ/425	–	SWI	Spear	Odysseus (Greek)	Haropas (Trojan)	–
100.	Λ/433–437	Thorax	SWI	Spear	Sokos (Trojan)	Odysseus (Greek)	Survival
101.	Λ/446–448	Thorax and back	SWI	Spear	Odysseus (Greek)	Sokos (Trojan)	Death
102.	Λ/489	–	SWI	–	Ajax (Greek)	Doryklos (Trojan)	Death
103.	Λ/490	–	–	–	Ajax (Greek)	Pandokos (Trojan)	–

(Continued)

Table 6.1 (Continued)

Case number	Rhapsody/ verses	Anatomical region	Type of trauma	Weapon	Attacker	Defender (injured)	Outcome
104.	Λ/491	–	SWI	–	Ajax (Greek)	Lysandros, Pyrasos, Pylartis (Trojans)	Death (x3)
105.	Λ/496–497	–	SWI	–	Ajax (Greek)	Trojan warriors and their horses	Deaths (Several)
106.	Λ/502–503	–	SWI and Blunt trauma	Spear and chariot	Hector (Trojan)	Greek warriors	Deaths (Several)
107.	Λ/505–507	Shoulder	SWI	Arrow	Paris (Trojan)	Machaon (Greek)	Survival
108.	Λ/576–579	Abdomen	SWI	Spear	Eyrypylos (Greek)	Apisaonas (Trojan)	Death
109.	Λ/581–584	Thigh	SWI	Arrow	Paris (Trojan)	Eyrypylos (Greek)	Survival
110.	M/116–117	–	SWI	Spear	Idomeneas (Greek)	Asios (Trojan)	Survival
111.	M/182–186	Head	SWI	Spear	Polypitis (Greek)	Damasos (Trojan)	Death
112.	M/187	–	–	–	Polypitis (Greek)	Ormenos, Pilonas (Trojans)	Deaths (x2)
113.	M/188–189	Abdomen	SWI	Spear	Leonteas (Greek)	Ippomachos (Trojan)	–

114.	M/190–194	–	SWI	Sword	Leonteas (Greek)	Antifatis, Menonas, Iamenos, Orestis (Trojans)	Deaths (x4)
115.	M/378–386	Head	Blunt trauma	Marble	Ajax (Greek)	Epikleas (Trojan)	Death
116.	M/387–389	Arm	SWI	Arrow	Teukros (Greek)	Glaukos (Trojan)	Survival
117.	M/392–395	–	SWI	Spear	Sarpidonas (Trojan)	Alkamas (Greek)	–
118.	N/170–180	Head	SWI	Spear	Teukros (Greek)	Imbrios (Trojan)	Death
119.	N/185–187	Thorax	SWI	Spear	Hector (Trojan)	Amphimachos (Greek)	Death
120.	N/210–214	Lower extremity	SWI ("χαλκῷ"—with bronze)	–	Trojan warrior	Idomeneas (Greek)	Survival
121.	N/370–375	Abdomen	SWI	Spear	Idomeneas (Greek)	Orthyoneas (Trojan)	Death
122.	N/384–391	Neck	SWI	Spear	Idomeneas (Greek)	Asios (Trojan)	Death
123.	N/395–399	Abdomen	SWI	Spear	Antolochos (Greek)	Trojan warrior (comrade of Asios)	–
124.	N/410–412	Abdomen	SWI	Spear	Diiphovos (Trojan)	Ypsinoras (Greek)	Survival
125.	N/438–443	Thorax	SWI	Spear	Idomeneas (Greek)	Alkathoos (Trojan)	Death
126.	N/506–509	Abdomen	SWI	Spear	Idomeneas (Greek)	Inomaos (Trojan)	Death
127.	N/517–520	Shoulder	SWI	Spear	Diiphovos (Trojan)	Askalathos (Greek)	–

(Continued)

Table 6.1 (Continued)

Case number	Rhapsody/ verses	Anatomical region	Type of trauma	Weapon	Attacker	Defender (injured)	Outcome
128.	N/527–533	Arm	SWI	Spear	Myrionis (Greek)	Diiphovos (Trojan)	Survival
129.	N/541–544	Neck	SWI	Spear	Aeneas (Trojan)	Afargias (Greek)	Death
130.	N/545–549	Back	SWI ("ἔκερσεν"—cut)	–	Antilochus (Greek)	Thoonas (Trojan)	–
131.	N/567–575	Abdomen	SWI	Spear	Myrionis (Greek)	Adamas (Trojan)	Death
132.	N/576–580	Head	SWI	Sword	Elenos (Trojan)	Diiporos (Greek)	Death
133.	N/593–600	Forearm	SWI	Spear	Menelaus (Greek)	Elenos (Trojan)	Survival
134.	N/610–620	Head	SWI	Sword	Menelaus (Greek)	Pissandros (Trojan)	Death
135.	N/650–654	Abdomen and Hip	SWI	Arrow	Myrionis (Greek)	Arpalionis (Trojan)	Death
136.	N/660–672	Head	SWI	Arrow	Paris (Trojan)	Eychinoras (Greek)	Death
137.	Ξ/410–413	Neck and thorax	Blunt trauma	Rock	Ajax (Greek)	Hector (Trojan)	Survival
138.	Ξ/442–447	Abdomen	SWI	Spear	Ajax of Oileas (Greek)	Satnios (Trojan)	–
139.	Ξ/449–452	Shoulder	SWI	Spear	Polydamas (Trojan)	Prothinoras (Greek)	–
140.	Ξ/459–469	Neck	SWI	Spear	Ajax (Greek)	Archelochos (Trojan)	Death
141.	Ξ/475–477	–	SWI	Spear	Akamas (Trojan)	Promachos (Greek)	Death

142.	Ξ/489–500	Head and neck	SWI	Spear	Pinelaos (Greek)	Ilioneas (Trojan)	Death
143.	Ξ/511	–	–	–	Ajax (Greek)	Yrtios (Trojan)	–
144.	Ξ/513	–	SWI (verb ἐξενάριξε (slay) used to describe injury)	–	Antilochus (Greek)	Mermeros, Philakis (Trojans)	Deaths (x2)
145.	Ξ/514	–	–	–	Myrionis (Greek)	Maris, Ippotionas (Trojans)	Deaths (x2)
146.	Ξ/515	–	–	–	Teukros (Greek)	Periphytos, Prothonas (Trojans)	–
147.	Ξ/516–519	Abdomen	SWI (“χαλκὸς“—bronze)	–	Menelaus or Agamemnonas (Greek)	Yperinonas (Trojan)	Death
148.	Ξ/520–521	–	–	–	Ajax of Oileas (Greek)	Several Trojan warriors	Deaths (Several)
149.	O/329–331	–		–	Hector (Trojan)	Stichios, Arkesilas (Greeks)	Deaths (x2)
150.	O/332	–	–	–	Aeneas (Trojan)	Medontas, Iasos (Greeks)	Deaths (x2)
151.	O/339	–	–	–	Polydamas (Trojan)	Mikistia (Greek)	Death
152.	O/339	–	–	–	Politis (Trojan)	Echios (Greek)	Death
153.	O/340	–	–	–	Aginoras (Trojan)	Klonios (Greek)	Death
154.	O/341–342	Shoulder	SWI	Arrow	Paris (Trojan)	Diochios (Greek)	–
155.	O/419–420	Thorax	SWI	Spear	Ajax (Greek)	Kalitoras (Greek)	–
156.	O/429–435	Head	SWI	Spear	Hector (Trojan)	Lykofronas (Greek)	Death

(Continued)

Table 6.1 (Continued)

Case number	Rhapsody/ verses	Anatomical region	Type of trauma	Weapon	Attacker	Defender (injured)	Outcome
157.	O/445	–	SWI	Arrow	Teukros (Greek)	Pissinoras (Trojan)	Death
158.	O/445–452	Neck	SWI	Arrow	Teukros (Greek)	Klitos (Trojan)	Death
159.	O/515	–	–	–	Hector (Trojan)	Perimidis (Greek)	Death
160.	O/516	–	–	–	Ajax (Greek)	Laodamas (Trojan)	Death
161.	O/518–519	–	–	–	Polydamas (Trojan)	Otos (Greek)	Death
162.	O/520–524	Thorax	SWI	Spear	Megis (Greek)	Krismos (Trojan)	Death
163.	O/535–538	Head	SWI	Spear	Megis (Greek)	Philidis (Trojan)	Survival
164.	O/540–543	Shoulder and thorax	SWI	Spear	Menelaus (Greek)	Philidis (Trojan)	Death
165.	O/572–578	Thorax	SWI	Spear	Antilochus (Greek)	Melanippos (Trojan)	Death
166.	O/638–652	Thorax	SWI	Spear	Hector (Trojan)	Periphitis (Greek)	Death
167.	O/742–746	–	SWI	Spear	Ajax (Greek)	Trojan warriors (x12)	Deaths (x12)
168.	Π/284–292	Shoulder	SWI	Spear	Patroclus (Greek)	Pyrechmis (Trojan)	Death
169.	Π/307–309	Thigh (Femoral fracture)	SWI	Spear	Patroclus (Greek)	Arilykos (Trojan)	–
170.	Π/311–312	Thorax	–	–	Menelaus (Greek)	Thoas (Trojan)	Death
171.	Π/313–316	Thigh	SWI	Spear	Philidis (Greek)	Amphiclos (Trojan)	Death

172.	Π/317–319	Abdomen	SWI	Spear	Antilochus (Greek)	Atymnios (Trojan)	Death
173.	Π/319–325	Shoulder (Humeral fracture)	SWI	Spear	Antilochus (Greek)	Maris (Trojan)	Death
174.	Π/330–334	Neck	SWI	Sword	Ajax of Oileas (Greek)	Kleovoulos (Trojan)	Death
175.	Π/337–341	Head	SWI	Sword	Pineleos (Greek)	Lykonas (Trojan)	Death
176.	Π/342–344	Shoulder	SWI	Spear	Myrionis (Greek)	Akamas (Trojan)	Death
177.	Π/345–350	Head	SWI	Spear	Idomeneas (Greek)	Eyrymantas (Trojan)	Death
178.	Π/377–379	–	Blunt trauma	Chariot	Patroclus (Greek)	Trojan warriors	Deaths (sever-al)
179.	Π/399–401	Thorax	SWI	Spear	Patroclus (Greek)	Pronoos (Trojan)	Death
180.	Π/401–410	Head	SWI	Spear	Patroclus (Greek)	Thestoras (Trojan)	Death
181.	Π/411–414	Head	Blunt trauma	Rock	Patroclus (Greek)	Eyrilaos (Trojan)	Death
182.	Π/415–418	–	–	–	Patroclus (Greek)	Eyrymantas, Amphoteras, Epaltis, Echios, Klipolemos, Pyris, Efvippos, Polymilos, Ifeas (Trojans)	Deaths (x9)
183.	Π/462–465	Abdomen	–	–	Patroclus (Greek)	Thrasymilos (Trojan)	–

(*Continued*)

Table 6.1 (Continued)

Case number	Rhapsody/ verses	Anatomical region	Type of trauma	Weapon	Attacker	Defender (injured)	Outcome
184.	Π/479–486	Thorax	SWI	Spear	Patroclus (Greek)	Sarpidonas (Trojan)	Death
185.	Π/570–580	Head	Blunt trauma	Rock	Hector (Trojan)	Epigeas (Greek)	Death
186.	Π/584–587	Neck	Blunt trauma	Rock	Patroclus (Greek)	Sthenalus (Trojan)	Death
187.	Π/593–598	Thorax	SWI	Spear	Glaukos (Trojan)	Vathyklis (Greek)	Death
188.	Π/603–607	Head	SWI	–	Myrionis (Greek)	Laogonos (Trojan)	Death
189.	Π/694–697	–	–	–	Patroclus (Greek)	Perimos, Elasos, Moulios, Pylartis, Epistoras, Melanippos, Autonoos, Echeklos, Adristos (Trojans)	Deaths (x9)
190.	Π/733–743	Head	Blunt trauma	Rock	Patroclus (Greek)	Kevrionis (Trojan)	Death
191.	Π/785	–	–	–	Patroclus (Greek)	Trojan warriors	Deaths (several)
192.	Π/787–795	Back	Blunt trauma	Hand	Apollo (God)	Patroclus (Greek)	Survival
193.	Π/808–815	Back	SWI	Spear	Efthorvos (Trojan)	Patroclus (Greek)	Survival
194.	Π/818–829	Abdomen	SWI	Spear	Hector (Trojan)	Patroclus (Greek)	Death

195.	P/45–49	Neck	SWI	Spear	Menelaus (Greek)	Efthorvos (Trojan)	Death
196.	P/293–298	Head	SWI	Spear	Ajax (Greek)	Ippothoos (Trojan)	Death
197.	P/306–311	Neck	SWI	Spear	Hector (Trojan)	Schedios (Greek)	Death
198.	P/312–315	Abdomen	SWI	–	Ajax (Greek)	Phorkinas (Trojan)	Death
199.	P/344–345	–	SWI	Spear	Aeneas (Trojan)	Liokritos (Greek)	–
200.	P/346–349	Abdomen	SWI	Spear	Lykomidis (Greek)	Apisaonas (Trojan)	–
201.	P/516–524	Abdomen	SWI	Spear	Aytomedas (Greek)	Aritios (Trojan)	Death
202.	P/575–579	Abdomen	SWI	Spear	Menelaus (Greek)	Podis (Trojan)	–
203.	P/597–600	Shoulder	SWI	Spear	Polydamas (Trojan)	Pinelaos (Greek)	Survival
204.	P/601–604	Wrist	–	–	Hector (Trojan)	Litos (Greek)	Survival
205.	P/610–618	Head	SWI	Spear	Hector (Trojan)	Kiranos (Greek)	Death
206.	Y/382–388	Head	SWI	Spear	Achilles (Greek)	Otrydidis (Trojan)	Death
207.	Y/395–400	Head	SWI	Spear	Achilles (Greek)	Dimoleos (Trojan)	Death
208.	Y/401–406	Back	SWI	Spear	Achilles (Greek)	Ippodamos (Trojan)	Death
209.	Y/407–418	Abdomen	SWI	Spear	Achilles (Greek)	Polydoros (Trojan)	Death
210.	Y/455–456	Neck	SWI	Spear	Achilles (Greek)	Dryopas (Trojan)	Death
211.	Y/457–459	Knee	SWI	Spear	Achilles (Greek)	Dimouchos (Trojan)	Survival
212.	Y/457–459	–	SWI	Sword	Achilles (Greek)	Dimouchos (Trojan)	Death

(*Continued*)

Table 6.1 (Continued)

Case number	Rhapsody/ verses	Anatomical region	Type of trauma	Weapon	Attacker	Defender (injured)	Outcome
213.	Y/460–462	–	SWI	Spear	Achilles (Greek)	Laogonos (Trojan)	Death
214.	Y/460–462	–	SWI	Sword	Achilles (Greek)	Dardanos (Trojan)	Death
215.	Y/463–471	Abdomen	SWI	Sword	Achilles (Greek)	Trojan warrior	Death
216.	Y/472–474	Head	SWI	Spear	Achilles (Greek)	Moulios (Trojan)	Death
217.	Y/474–477	Head	SWI	Sword	Achilles (Greek)	Echeklos (Trojan)	Death
218.	Y/478–481	Elbow	SWI	Spear	Achilles (Greek)	Defkalionas (Trojan)	Survival
219.	Y/481–483	Neck	SWI	Sword	Achilles (Greek)	Defkalionas (Trojan)	Death
220.	Y/484–487	Abdomen and back	SWI	Spear	Achilles (Greek)	Rigmos (Trojan)	Death
221.	Y/487–489	Back	SWI	Spear	Achilles (Greek)	Arithos (Trojan)	Death
222.	Φ/17–21	–	SWI	Sword	Achilles (Greek)	Trojan warriors	Deaths (several)
223.	Φ/115–120	Neck	SWI	Sword	Achilles (Greek)	Lykaonas (Trojan)	Death
224.	Φ/164–168	Hand	SWI	Spear	Asteropios (Trojan)	Achilles (Greek)	Survival
225.	Φ/179–182	Abdomen	SWI	Sword	Achilles (Greek)	Asteropios (Trojan)	Death
226.	Φ/209–210	–	SWI	Sword	Achilles (Greek)	Mydonas, Thersilochus, Astypylos, Thrasios,	Deaths (x7)

						Ofelestis, Mnysios, Aenios (Trojans)	
227.	Φ/403–408	Neck	Blunt trauma	Rock	Athina (Goddess)	Ares (God)	Survival
228.	Φ/423–426	Thorax	Blunt trauma	Hand	Athena (Goddess)	Aphrodite (Goddess)	Survival
229.	X/325–327	Neck	SWI	Spear	Achilles (Greek)	Hector (Trojan)	Death
230.	Ψ/391–397	–	Blunt trauma and lacerations	fall off chariot during funeral games held for Patrocl-us	–	Evmilos (Greek)	Survival
231.	Ψ/677–679	Head	Blunt trauma (sports injury during the funeral games)	Fist	Epios (Greek)	Evrialos (Greek)	Survival
232.	Ψ/708–732	–	Blunt trauma (sports injury during the funeral games)	Fist	Odysseus (Greek)	Ajax (Greek)	Survival
233.	Ψ/708–732	–	Blunt trauma (sports injury during the funeral games)	Fist	Ajax (Greek)	Odysseus (Greek)	Survival
234.	Ψ/828	–	–	–	Achilles (Greek)	Ietionas (Trojan)	Death

SWI, Sharp weapon injury.

Table 6.2 Trauma registry of all wounds that are mentioned in the *Iliad* but did not occur during the Trojan War (T.W.: Trojan War).

No	Rhapsody/ verses	Anatomical region	Type of trauma	Weapon	Attacker	Defender (injured)	Outcome
1.	Z/134–135	–	–	–	Lykourgos (King of Thrace, Greece)	Nurses of Dionysos (Greeks)	Survival (not in T.W. past mythological events)
2.	Z/139	Eyes	Blindness	–	Zeus (God)	Lykourgos (King of Thrace, Greece)	Survival (not in T.W. past mythological events)
3.	Z/183	–	–	–	Vellerefontis (Greek hero of the past)	Chimaira (Monster)	Death (not in T.W. past mythological events)
4.	Z/186	–	–	–	Vellerefontis (Greek hero of the past)	Amazons	Death (not in T.W. past mythological events)
5.	Z/189–190	–	–	–	Vellerefontis (Greek hero of the past)	Men from Lykia (Greeks)	Death (not in T.W. past mythological events)
6.	Z/203–204	–	–	–	Mars (God)	Isandros (son of Vellerefontis, Greek)	Death (not in T.W. past mythological events)
7.	Z/414–416	–	SWI	Sword	Achilles (Greek)	Ietionas (father of Andromcahi) (T)	Death (not in T.W. past events)
8.	Z/421–423	–	SWI	Sword	Achilles (Greek)	Adromachi's (wife of Hector) 7 siblings (T)	Death (x7) (not in T.W. past events)
9.	Z/428	–	SWI	Arrow	Artemis (God)	Andromachi's (wife of Hector) mother (T)	Death (not in T.W. past events)

10.	H/142–145	Back	SWI	Spear	Lykourgos (King of Thrace, Greece)	Arithos (Greek)	Death (not in T.W. past events)
11.	H/148–156	–	–	–	Nestoras (Greek)	Nevrethalios (comrade of Lykourgos, King of Thrace, Greek)	Death (Not in T.W. past events)
12.	Λ/673–675	–	SWI	Spear	Nestoras (Greek)	Itimos (son of Ypeirochos, from Ilida, Greek city state)	Death (not in the T.W. past events—battle between Pylos and Ilida)
13.	Λ/737–743		SWI	Spear	Nestoras (Greek)	Moulios (Ilida—Greek city state)	Death (not in the T.W. past events—battle between Pylos and Ilida)
14.	Λ/747–749	–	–	–	Nestoras (Greek)	Multiple warriors (Ilida—Greek city state)	Deaths (not in the T.W. past events—battle between Pylos and Ilida)
15.	Λ/759	–	–	–	Nestoras (Greek)	Warrior (Ilida—Greek city state)	Death (not in the T.W. past events)

Cases **highlighted in yellow** are those not explicitly cataloged in other studies.

(Chicco & Tebala, 2021; Nomikos, 2018). Kömürcü et al. detected 103 musculoskeletal injuries across 81 combats (Kömürcü et al., 2014). Mylonas et al. identified 48 cranio-maxillofacial injuries in the *Iliad* (Mylonas et al., 2008). This variance in reported figures can be attributed to several factors: first, investigators relied on different translations of the *Iliad*, of which more than 200 versions have been published and acquiring the original Greek text without omissions or securing an accurate version posed challenges for many translators. Second, most investigators of these injuries are physicians rather than philologists or historians, potentially influencing their description and interpretation, since it is important to follow the story of the poem to gather all available information from various points in the text. Furthermore, battle descriptions might focus on a single injury type, while extensively detailing the injury's scope (involving the deconstruction of the original text), potentially leading to the enumeration of multiple injuries. Finally, some investigators may have not included non-fatal injuries, or traumas described between gods or those not occurring at the siege of Troy.

The purpose of this chapter is to provide a complete trauma registry of the *Iliad*. Hence, the original text of this Homeric poem was meticulously studied. It is of note that Homer did not intend to write a medical manuscript but a poem narrating epic battles and heroic events of the past. Nevertheless, the poem's verses contain information about the nature of traumas and anatomy or trauma management. The present listing represents the most thorough one, so far in the literature and it analyzes each case separately.

Regarding the mortality of these injuries, SWIs by a spear, sword or arrow exhibited a high mortality rate of almost 80%. Furthermore, injuries located at the head and neck, as well as the thorax and abdomen depicted much higher mortality rates than the ones located at the upper or lower extremity.

Furthermore, Homer vividly describes some instances of fractures, including a broken femur (Case 169 in Table 6.1) and a broken humerus (Case 173 in Table 6.1). The protective role of the warriors' armors and especially the helmets is also highlighted. For example, in Case 94 of Table 6.1 Hector's helmet saved him from a lethal injury and led to a concussion, while in Case 163, Table 6.1 another Trojan warrior (Philidis) survived a similar injury. In another instance, the shield and armor of Odysseus saved him from a thorax injury by spear (Case 100, Table 6.1). On the other hand, in Case 115, Table 6.1 the Greek warrior Ajax shattered the helmet of the Trojan warrior Epikleas using a piece of marble. Once the helmet was broken, it failed to provide protection, leading to Epikleas's death.

Diomedes and Achilles represent the main "attackers" since over 50 deaths are attributed to them. The wrath of Achilles is mainly depicted in Rhapsody Y, after he finds out that his trusted comrade Patroclus has been killed on the battlefield (see Fig. 6.1).

There are also three instances, Cases 231–233 of Table 6.1, that describe sports injuries that occurred during the boxing fights at the Greek camp honoring the dead warrior Patroclus. It is also of note that probably a case of a brachial plexus injury is also described (Case 74 in Table 6.1) where Hector hit the Greek warrior Teukros

Figure 6.1 Achilles mourning the death of Patroclus. Achilles cutting off a lock of his hair in mourning over the body of Patroclus. Engraving by P. Lombard, 1660, after A. van Diepenbeeck.
Source: Data from Wellcome Collection. https://wellcomecollection.org/works/z3vk7nk8.

with a rock at the shoulder, which led to neuropraxia of the whole arm. (See also Chapter 15 for more on this case.)

The *Iliad* has long been studied to gain insights into battle injuries, anatomical knowledge, and trauma management in Ancient Greece (Chicco & Tebala, 2021; Koutserimpas & Samonis, 2017; Koutserimpas et al., 2017; Nomikos, 2018). In the present chapter, a complete trauma registry of the *Iliad* is provided, based on the original text. It seems that Homer understood the nature of wounds, and which ones were lethal, since vital organs could be damaged. The head and neck, as well as the trunk regions, were much more likely than other body regions to be involved in fatal injuries, while most deaths were *in situ*.

Conclusions

Within the *Iliad* and *Odyssey* (see Chapter 7) lie significant depictions of battles and injuries, offering insights into the nature of wounds and their consequences during that era. Despite the considerable advancements in medical sciences in recent decades, examining the nature of traumas and their outcomes during the early stages of Western civilization contributes to comprehending the trajectory and development of medicine across history.

References

Apostolakis, E., Apostolaki, G., Apostolaki, M., & Chorti, M. (2010). The reported thoracic injuries in Homer's *Iliad*. *Journal of Cardiothoracic Surgery, 5*(1). https://doi.org/10.1186/1749-8090-5-114, http://www.cardiothoracicsurgery.org/content/5/1/114 Greece.

Chicco, M., & Tebala, G. D. (2021). War trauma in Homer's *Iliad*: A trauma registry perspective. *European Journal of Trauma and Emergency Surgery, 47*(3), 773–778. https://doi.org/10.1007/s00068-020-01365-6, http://link.springer.com/journal/68.

Fröhlich, H. (1879). *Die Militärmedizin Homer's (The military medicine of Homer)*. Enke.

Homer. (n.d.). *The Iliad*.

Kakridis I: *Greek* mythology (1st ed.). (1986).

Koutserimpas, C., Alpantaki, K., & Samonis, G. (2017). Trauma management in Homer's *Iliad*. *International Wound Journal, 14*(4), 682–684. https://doi.org/10.1111/iwj.12672, http://onlinelibrary.wiley.com/journal/10.1111/(ISSN)1742-481X.

Koutserimpas, C., & Samonis, G. (2017). Letter to the editor: Medical structures during war in the Homeric era: Paramedics and 'expedition' hospitals. *Journal of Wound Care, 26*(3), 152–153. https://doi.org/10.12968/jowc.2017.26.3.152b.

Kömürcü, E., Tok, F., Şimşek, A., & Özçakar, L. (2014). Musculoskeletal injuries in Homer's *Iliad*: The war of Troy revisited. *American Journal of Physical Medicine and Rehabilitation, 93*(4), 335–341. https://doi.org/10.1097/PHM.0000000000000051, http://www.lww.com/store/products?0894-9115.

Mylonas, A. I., Tzerbos, F. H., Eftychiadis, A. C., & Papadopoulou, E. C. (2008). Craniomaxillofacial injuries in Homer's *Iliad*. *Journal of Cranio-Maxillo-Facial Surgery, 36*(1), 1–7. https://doi.org/10.1016/j.jcms.2007.07.010.

Nomikos, N. (2018). Injuries in the Greek epics of Homer. *Chinese Journal of Traumatology - English Edition, 21*(2), 109–112. https://doi.org/10.1016/j.cjtee.2017.09.005, http://www.elsevier.com/wps/find/journaldescription.cws_home/714951/description#description.

Passas, J. (1960). *New encyclopedic dictionary*.

Santos, G. H. (2000). Chest trauma during the Battle of Troy: Ancient warfare and chest trauma. *Annals of Thoracic Surgery, 69*(4), 1285–1287. https://doi.org/10.1016/S0003-4975(00)01158-9, http://www.elsevier.com/locate/athoracsur.

A trauma registry of Homer's *Odyssey*

7

Christos Koutserimpas
School of Rehabilitation Health Sciences, University of Patras, Greece

Introduction

The *Odyssey*, an epic narrative spanning 24 rhapsodies (books), conventionally attributed to the ancient Greek poet Homer, recounts the adventures of Odysseus, the ruler of Ithaca (Homer, no date). Following the conclusion of the Trojan War, Odysseus embarks on a perilous journey lasting a decade, though the poem's events unfold over a mere six weeks. His primary objective is to return home. Upon his arrival, he finds recognition solely from his loyal dog and a nurse (Homer, no date; Miranda et al., 2018).

Guided by his son Telemachus, Odysseus takes decisive action against the persistent mnesteres (suitors) who have sought the hand of his faithful wife, Penelope. Additionally, he confronts and punishes several of Penelope's maids who had aligned themselves with the mnesteres. Through these trials, Odysseus restores order to his kingdom and reinstates himself as the rightful ruler of the island of Ithaca (Homer, no date).

Homer's *Iliad* has undergone thorough examination regarding trauma cases (Chicco & Tebala, 2021; Kömürcü et al., 2014; Nomikos, 2018) and their management (Koutserimpas & Samonis, 2017b; Koutserimpas et al., 2017); however, the same level of scrutiny has not been applied to the *Odyssey*. There, thus far, has been a list of injuries in the *Odyssey* published (Koutserimpas & Samonis, 2016), as well as a discussion of head and neck injuries in the *Odyssey* (Koutserimpas & Samonis, 2017a; Stathopoulos et al., 2016; Stathopoulos et al., 2017).

The present chapter focuses on traumas and their nature in the *Odyssey* by Homer, aiming to systematically document all forms of injuries depicted in this epic poem.

Methods

Each rhapsody and verse of Homer's original text in the *Odyssey* underwent meticulous examination to identify every instance of trauma. Scrutiny was applied to discern details such as the type and site of the injury, the assailant, the victim, the weapon utilized, and the ultimate outcome of the wound. Relevant references from the PubMed database were also consulted.

Results

Table 7.1 provides a thorough listing of all traumas described in Homer's *Odyssey*.

Medicine in Homer. DOI: https://doi.org/10.1016/B978-0-443-32978-4.00012-2

Table 7.1 The complete trauma registry of the *Odyssey* (G = Greek; m = mnester (suitor)).

Case number	Rhapsody/ verses	Anatomical region	Type of trauma	Weapon	Attacker	Defender (injured)	Outcome
1	E/128	Whole body	Electric shock	Thunder	Zeus (God)	Iasios (G)	Death
2	E/123–124	–	SWI	Arrow	Artemis (God)	Orion (G)	Death
3	E/428–435	Upper extremity	Burn	Wave, rocks	Poseidon (God)	Odysseus (G)	Survival
4	Θ/528	Back	SWI	Spear	Greek warriors	Trojan woman	Survival
5	I/287–289	Skull	Fracture—fall (blunt trauma)	–	Polyphemus (cyclops)	Odysseus' comrades (×2)	Death (×2)
6	I/381–390	Eye	SWI—blindness	Wooden stake	Odysseus (G)	Polyphemus (cyclops)	Survival
7	K/123–124	–	SWI	Wooden stake	Laestrygones (man-eating giant)	Odysseus' comrades (G)	Deaths (several)
8	K/556–560	Cervical spine	Blunt trauma—fall	–	–	Elpenor (G)	Death (see Chapter 15 for discussion of Elpenor's injury.)
9	Λ/518–521	–	SWI	–	Neoptolemus (G)	Eurylipus (G)	Death
10	M/411–414	Skull	Fracture (blunt trauma)	Ship's mast	–	Eurylochus (G)	Death
11	M/415–419	Whole body	Electric shock	Thunder	Zeus (God)	Odysseus' comrades (G)	Deaths (several)
12	N/268–269	–	SWI	Spear	Odysseus (G)	Orsilochus (G)	Death
13	P/462–464	Shoulder	Blunt trauma	Stool	Antinous (m)	Odysseus (G)	Survival

14	Σ/95–98	Shoulder	Blunt trauma	Punch	Irus (a beggar in Ithaca who is willing to run messages for the mnesteres)	Odysseus (G)	Survival
15	Σ/95–98	Jaw and cervical spine	Blunt trauma	Punch	Odysseus (G)	Irus (a beggar in Ithaca who is willing to run messages for the mnesteres)	Survival
16	Σ/396–398	Hand	Blunt trauma	Punch	Eurymachus (m)	Odysseus' servant (G)	Survival
17	T/449–451	Knee	Laceration	Bite	Wild Boar	Odysseus (G)	Survival
18	Φ/299–302	Nose and Ears	SWI	Knife	Latithes (mythological creature similar to centaur)	Euritionas (m)	Survival
19	X/15–19	Throat and cervical spine	SWI	Arrow	Odysseus (G)	Antinous (m)	Death
20	X/81–84	Thorax	SWI	Arrow	Odysseus (G)	Eurymachus (m)	Death
21	X/91–92	Thorax and back	SWI	Spear	Telemachus (G)	Amphinomus (m)	Death
22	X/187–191	Upper and lower extremities	Hanging	Rope	Eumaios and Menelaus' confident (G)	Melanthius (m)	Survival
23	X/277–278	Upper extremity	SWI	Spear	Amphimedon (m)	Telemachus (G)	Survival
24	X/279–280	Shoulder	SWI	Spear	Ktesipus (m)	Eumaios (G)	Survival
25	X/283–284	–	SWI	Spear	Odysseus (G)	Eurydamas (m)	Death

(Continued)

Table 7.1 (Continued)

Case number	Rhapsody/verses	Anatomical region	Type of trauma	Weapon	Attacker	Defender (injured)	Outcome
26	X/283–284	–	SWI	Spear	Telemachus (G)	Amphimedon (m)	Death
27	X/283–284	–	SWI	Spear	Eumaios (G)	Polivos (m)	Death
28	X/285	Thorax	SWI	Spear	Menelaus' confidant (G)	Ktesipus (m)	Death
29	X/292–293	–	SWI	Spear	Odysseus (G)	Agelaos (m)	–
30	X/294–295	Groin area	SWI	Spear	Telemachus (G)	Leocritus (m)	Death
31	X/308–309	Head	Blunt trauma	Punch	Odysseus and comrades (G)	Mnesteres	Deaths (several)
32	X/362–369	Cervical spine	SWI	Sword	Odysseus (G)	Leiodes (m)	Death
33	X/474–478	Nose and ears	SWI	Knife	Telemachus (G)	Melanthius (m)	Death
34	Ω/522–524	Head	SWI	Spear	Laertes (G)	Eupidis (m)	Death

In particular, a total of 34 injuries were identified. The majority of them are located in rhapsody X (15 cases; 44.1%) which contains the description of Odysseus's revenge against the mnesteres.

Regarding the anatomical region, most described injuries were at the head and neck (11; 32.4%), followed by the upper extremity (6; 17.6%) and trunk (back, thorax, and abdomen; 4 cases, 11.8%). There are also two lower limb and groin area injuries, as well as two electrocutions involving the whole body (5.9%). Finally, there is 1 (2.9%) injury including both upper and lower extremities, while 8 (23.5%) of them do not specify the injured anatomical area.

A total of 21 (61.8%) injuries could be characterized as sharp weapon injuries (SWIs), including arrows, spears, wooden stakes, knives, and swords. Additionally, there are 8 (23.5%) blunt traumas, 2 (5.9%) electrocutions, 1 burn injury, 1 laceration (attack by an animal), and 1 hanging with rope (2.9% each).

Regarding the outcome of these injuries, most of them (24 cases; 70.6%) were fatal. SWIs were fatal in most cases (16 out of 21; 76.2%), while blunt traumas had 50% mortality. Both cases of electrocution were also fatal. Trunk injuries revealed 83.3% mortality, while head and neck 72.7%.

Discussion

The *Odyssey* unfolds with a narrative structure that defies linearity when compared to the *Iliad*. Rather than beginning at the tale's inception, readers are thrust into the middle, piecing together past events through Odysseus's recounted experiences (Homer, no date; Miranda et al., 2018).

Homer's *Odyssey* recounts the 10-year journey of Odysseus as he endeavors to return home following the Trojan War. Throughout his travels, Odysseus confronts mythical creatures and endures the wrath of the gods, while his wife Penelope and son Telemachus fend off numerous suitors seeking Penelope's hand and Ithaca's throne until Odysseus' return. The epic poem concludes with Odysseus proving his identity in a contest, vanquishing the suitors, and reclaiming the throne of Ithaca (Homer, no date).

The *Iliad* and the *Odyssey* offer significant insights into battles and the resulting traumas, providing valuable information on the nature of wounds and their consequences during that era. Trauma and its management in the *Iliad* have been meticulously studied and discussed (Kömürcü et al., 2014; Koutserimpas & Samonis, 2017b; Koutserimpas & Samonis, 2018; Nomikos, 2018). However, the same cannot be said for trauma cases in the *Odyssey*. The present chapter provided a complete trauma registry of this poem, analyzing the anatomical distribution of the described injuries, the type of traumas, the used weapon of each case, the characters involved, as well as the final outcome. A total of 34 injuries were recorded. It is of note that there are another two injuries mentioned in the poem as news or old stories that do not happen in the timeframe of the *Odyssey*. The story of Oedipous killing his father with sharp weapon is mentioned in rhapsody λ, while the story of Aegistos killing Agamemnonas (both Greek warriors) is also provided in the same rhapsody.

Figure 7.1 The cyclops Polyphemus attacking the comrades of Odysseus. Polyphemus attacking the companions of Odysseus and their ships, or the companions and ships of Aeneas. Etching with engraving.
Source: Data from Wellcome Collection. https://wellcomecollection.org/works/dmmzwwa6.

Despite the advancements in medical sciences in recent decades, examining the nature of traumas and their outcomes in ancient times aids in comprehending the trajectory and development of medicine over the centuries. In particular, it seems that the anatomical area of the injury, as well as the weapon that was used, contributed to the final outcome. SWIs inflicted by spears, stakes, arrows, knives, and swords exhibited a high mortality rate of about 76%. On the contrary, blunt traumas were fatal in only half of the cases, most of which included falls from a great height or extremely heavy objects falling on the protagonists, such as in Case 10 of Table 7.1 when a ship's mast fell on Eurylochus resulting in his death. Another

Figure 7.2 The blinding the cyclops Polyphemus. Odysseus and his men blinding the cyclops Polyphemus with a sharpened stake. Etching by Taylor, ca. 1790.
Source: Data from Wellcome Collection. https://wellcomecollection.org/works/jcfu9jht.

example is the "tossing" of two of Odysseus' comrades from Polyphemus' attack resulting in skull fractures and their death (Case 5 of Table 7.1 and Fig. 7.1).

Injuries in different anatomical areas were also associated with variable mortality rates. For instance, injuries of the upper extremities were not considered fatal, showcasing that they could be treated with the medical knowledge of that time. On the other hand, injuries to the head and neck, the abdomen, and thorax exhibited very high mortality rates, up to approximately 83%. This shows that the nature of these wounds is much more complex and difficult to manage when compared to the extremities.

The most renowned injury that is described in the *Odyssey* is the blinding of the Cyclops Polyphemus by Odysseus, depicted in Fig. 7.2 and Case 6 of Table 7.1.

In this instance, Odysseus did not want to kill the Cyclops Polyphemus because he had trapped him and his comrades in his cave with a huge rock that none could move besides the giant Cyclops. His comrades sharpened a wooden stake that was also put into the fire and, as soon as Polyphemus was drunk and fell asleep, they blinded him. During their conversation, Odysseus told Polyphemus that his name was «Οὖτις» meaning "Nobody." Hence, when Polyphemus was blinded and screamed for help to his fellow Cyclopes saying: "ὦ φίλοι, Οὖτίς με κτείνει δόλῳ οὐδὲ βίηφιν" ("Oh friends, Nobody is killing me by trickery, not by force") they replied if nobody blinded you, why do you yell? After escaping from the Cyclops, tied under the bellies of Polyphemus' sheep, Odysseus, while sailing away, screamed to the giant his true name, showing his arrogance. Polyphemus then, prayed to his father, God Poseidon, to punish Odysseus and that led to the wandering that lasted 10 whole years.

The injuries described in the *Odyssey* share many similarities with those in the *Iliad*. Both texts reveal that SWIs and traumas to the head, neck, thorax, and abdomen had a significantly higher likelihood of resulting in death compared to injuries to the extremities. The *Iliad*, with its extensive focus on continuous battles, provides more detailed information about the nature of wounds and their outcomes. However, both epic poems offer valuable insights into the types of injuries sustained and the trauma care practices of the time.

Conclusions

The trauma registry of the *Odyssey* completes the one from the *Iliad*, presenting all traumas in these two epic poems. It is fascinating to comprehend what medicine meant at that time. These epics provide valuable insights into the nature of wounds, as well as their outcomes.

References

Chicco, M., & Tebala, G. D. (2021). War trauma in Homer's *Iliad*: A trauma registry perspective. *European Journal of Trauma and Emergency Surgery, 47*(3), 773–778. https://doi.org/10.1007/s00068-020-01365-6, http://link.springer.com/journal/68.

Homer. The Odyssey.

Koutserimpas, C., Alpantaki, K., & Samonis, G. (2017). Trauma management in Homer's *Iliad*. *International Wound Journal, 14*(4), 682–684. https://doi.org/10.1111/iwj.12672, http://onlinelibrary.wiley.com/journal/10.1111/(ISSN)1742-481X.

Koutserimpas, C., & Samonis, G. (2016). Letters. *Journal of Wound Care, 25*(9), 553–554. https://doi.org/10.12968/jowc.2016.25.9.553.

Koutserimpas, C., & Samonis, G. (2017a). Re: Stathopoulos et al. Injuries to the head and neck in Homer's Odyssey. *British Journal of Oral and Maxillofacial Surgery, 55*(2), 217. https://doi.org/10.1016/j.bjoms.2016.06.017, http://www.elsevier-international.com/journals/bjom/.

Koutserimpas, C., & Samonis, G. (2018). Machaon: The first trauma surgeon in Western history? *Journal of Wound Care, 27*(10), 659–661. https://doi.org/10.12968/jowc.2018.27.10.659, http://www.internurse.com/cgi-bin/go.pl/library/issues.html?journal_uid=38.

Koutserimpas, C., & Samonis, G. (2017b). Letter to the editor: Medical structures during war in the Homeric era: Paramedics and 'expedition' hospitals. *Journal of Wound Care, 26*(3), 152–153. https://doi.org/10.12968/jowc.2017.26.3.152b.

Kömürcü, E., Tok, F., Şimşek, A., & Özçakar, L. (2014). Musculoskeletal injuries in Homer's *Iliad*: The war of Troy revisited. *American Journal of Physical Medicine and Rehabilitation, 93*(4), 335–341. https://doi.org/10.1097/PHM.0000000000000051, http://www.lww.com/store/products?0894-9115.

Miranda, P. J., Baptista, & Pinto, S. (2018). The Odyssey's mythological network. *PLoS One, 13*.

Nomikos, N. (2018). Injuries in the Greek epics of Homer. *Chinese Journal of Traumatology - English Edition, 21*(2), 109–112. https://doi.org/10.1016/j.cjtee.2017.09.005, http://www.elsevier.com/wps/find/journaldescription.cws_home/714951/description#description.

Stathopoulos, P., Ghaly, G. A., & Azari, A. (2016). Injuries to the head and neck in Homer's Odyssey. *British Journal of Oral and Maxillofacial Surgery, 54*(6), 717–718. https://doi.org/10.1016/j.bjoms.2015.10.018.

Stathopoulos, P., Ghaly, G. A., & Azari, A. (2017). Response to the Letter to the Editor re: Stathopoulos et al. Injuries to the head and neck in Homer's Odyssey. *The British Journal of Oral & Maxillofacial Surgery, 55*(2). https://doi.org/10.1016/j.bjoms.2016.06.025.

Trauma medicine in Homer

8

Christos Koutserimpas
School of Rehabilitation Health Sciences, University of Patras, Greece

Introduction

The *Iliad*, attributed to Homer, stands as one of ancient Greece's key epic poems. It's a portrayal of the Trojan War's concluding phase, a decade-long siege of Troy by a coalition of Mycenaean Greek states. Divided into 24 books (rhapsodies) and composed in dactylic hexameter, this work comprises about 15,693 verses in its widely accepted form. Focusing on crucial events during the war's final stages, it highlights the intense conflict between King Agamemnon and the famed warrior Achilles. Considered a pivotal part of the Epic Cycle, this classic is often hailed as one of the earliest and most influential pieces of European literature, still widely appreciated by modern readers.

The *Iliad* and *Odyssey*, composed between 800 BC and 700 BC, chronicle heroic events from the distant past, approximately five to six centuries prior. They also provide insights into the daily life and military practices of Homer's time. These epic works extensively mention medical and surgical details, showcasing a wealth of anatomical knowledge derived from experiences with traumas and fractures. Within these poems, blood holds a central role as the primary component of the human body, its loss equating to death. The soul is depicted as intangible, the essence of life departing the body with the final exhalation. The therapeutic approaches described are empirical, occasionally intertwined with perceptions of divine intervention.

The considerable depth of medical insight in these epics has sparked speculation that Homer himself might have been a military physician (Godquin, 1990) or descended from the Asclepiads, a group of lay healers. However, concrete evidence supporting these claims remains elusive. Additionally, the *Iliad* frequently references two prominent "fighting doctors," Machaon and Podalirius, serving as chief medical officers for the Greek forces during that time.

In the present chapter, the concept of trauma management and treatment is discussed through meticulous study of Homer's *Iliad*, as well as relevant literature.

Methods

Careful analysis involved the thorough study of the original texts of Homer's *Iliad* and *Odyssey*, alongside the comprehensive exploration of related myths from Ioannis Kakridis's five-volume work on Ancient Greek Mythology (Kakridis, 1986). This investigation extended to relevant literature and references, utilizing resources from

Medicine in Homer. DOI: https://doi.org/10.1016/B978-0-443-32978-4.00021-3

PubMed and Medline databases. The aim was to extract detailed insights into trauma management and the healthcare systems and structures prevalent during that historical era.

Treatment of injured warriors with rest

Table 8.1 summarizes all trauma cases mentioned in Homer's *Iliad* that "rest" served as the primary treatment. A total of 6 cases were described (Koutserimpas et al., 2017). A fall from height was treated with bed rest with a favorable outcome. Furthermore, a snake bite, a collarbone fracture, and penetrating injuries of the shoulder and abdomen, as well as a jaw punch are the injuries that were successfully managed with "rest." It is of note that in three of these cases, the immediate transport of the injured warriors is also described (cases 3–5 of Table 8.1). These instances highlight the significance of rest as a beneficial form of treatment for various injuries in the epic narrative of the *Iliad*.

Nonsurgical treatment of traumas

Table 8.2 summarizes all trauma cases in *Iliad* that "non-surgical" treatment was described. This category comprises three cases, characterized by the application of herbs known for their soothing and pain-relieving properties (Koutserimpas et al., 2017).

Surgical interventions

A total of 12 cases describing some form of surgical intervention and/or wound care are described in Homer's *Iliad* (see Table 8.3) (Koutserimpas et al., 2017). There are many cases of arrow and/or spear extractions, most of them performed by warriors (cases 1, 2, 4, and 6–9 of Table 8.3). Wound care and cleaning have been also reported in many instances (cases 3, 5, 8, 9, 10, and 11 of Table 8.3), while case 12 describes the application of pressure for stoppage of bleeding. The most commonly affected anatomical site was the upper extremity (6 cases; 50%), followed by lower extremities, as well as trunk and thorax injuries (3 each; 25%). Extraction of arrows was performed in 7 cases (58.3%) and of spears in 4 (33.3%). Cases 2 and 4 are illustrated in Figs. 8.1 and 8.2, respectively.

Healthcare structures in the Trojan War

Numerous Greek warriors possessed rudimentary wound care abilities, yet Machaon stood out as the principal medical officer and a specialized expert in managing trauma, overseeing the treatment of countless injured soldiers (Koutserimpas & Samonis, 2018). His tent served as a counterpart to a modern deployed military expedition hospital. Machaon's therapeutic approach encompassed herbal remedies, ointments, natural concoctions, and minor surgical procedures like extractions (Koutserimpas & Samonis, 2017).

Table 8.1 Trauma cases treated with rest in Homer's *Iliad*.

Case number	Case number. book/verses	Type of treatment	Type of trauma	Injured	Care provider	Final outcome following treatment
1	Α/594	Rest	Fall from height	Hephaestus (God)	not stated	Survival
2	Β/721–725	Rest	Snake bite	Philoctetes (Greek)	Philoctetes	Survival
3	Θ/332–334	Rest (transport)	Collar bone injury from a stone	Teucer (Greek)	Mecisteus and Alastor (Greeks)	Survival
4	Λ/510–515	Rest (transport)	Shoulder injury from an arrow	Machaon (Greek)	Nestor (Greek)	Survival
5	Ν/421–423	Rest (transport)	Abdomen injury from a spear	Hypsenor (Greek)	Mecisteus and Alastor (Greeks)	Death
6	Ψ/695–698	Rest	Punch of jaw	Euryalus (Greek)	other Greek warriors	Survival

Table 8.2 Trauma cases treated with non-surgical ways.

Case number	Case number. book/verses	Type of treatment	Type of trauma	Injured/care provider	Final outcome following treatment
1	E/401–402	Application of soothing herbs	Shoulder injury from an arrow	Hades (God)/Paeeon (God)	Survival
2	E/447	Divine intervention	Hip injury from a stone	Aeneas (Trojan)/Athena and Leto (Godesses)	Survival
3	E/900–904	Pain-killing herb treatment	Abdomen injury from a spear	Ares (God)/Paeeon (God)	Survival

Table 8.3 Trauma cases in Homer's *Iliad* that some form of surgical intervention was performed.

Case number	Book/ verses for treatment	Text	Type of treatment	Type of trauma	Injured/care provider	Final outcome following treatment
1	Injury: Δ/ 134–140 Treatment: Δ/212–219	On the clasped belt lighted the bitter arrow, [135] and through the belt richly dight was it driven, and clean through the curiously wrought corselet did it force its way, and through the taslet which he wore, a screen for his flesh and a barrier against darts, wherein was his chiefest defense; yet even through this did it speed. So the arrow grazed the outermost flesh of the warrior [Menelaus], [140] and forthwith the dark blood flowed from the wound. the godlike hero [Machaon] came and stood in their midst, and straightway drew forth the arrow from the clasped belt; and as it was drawn forth the sharp barbs were broken backwards. [215] And he loosed the flashing belt and the kilt beneath and the taslet that the coppersmiths fashioned. But when he saw the wound where the bitter arrow had lighted, he sucked out the	Arrow extraction, wound care, and cleaning, application of drugs (see Fig. 11.3)	Abdomen injury from an arrow	Menelaus (Greek)/ Machaon (Greek)	Survival

(Continued)

Table 8.3 (Continued)

Case number	Book/ verses for treatment	Text	Type of treatment	Type of trauma	Injured/care provider	Final outcome following treatment
		blood, and with sure knowledge spread thereon soothing simples, which of old Cheiron had given to his father with kindly thought.				
2	Injury: E/ 95–100 Treatment: E/111–113	But when the glorious son of Lycaon [Pandaros (Trojan)] was ware of him as he raged across the plain and drove the battalions in rout before him, forthwith he bent against the son of Tydeus [Diomed] his curved bow, and with sure aim smote him as he rushed onwards upon the right shoulder on the plate of his corselet; through this sped the bitter arrow [100] and held straight on its way, and the corselet was spattered with blood. and Sthenelus leapt from his chariot to the ground, and stood beside him [Diomed], and drew forth the swift arrow clean through his shoulder; and the blood spurted up through the pliant tunic.	Arrow extraction (see Fig. 8.1)	Shoulder injury from an arrow	Diomed (Greek)/ Sthenelus (Greek)	Survival

3	E/416–417		Wound care and cleaning	Hand injury from a spear	Venus (God)/ Dione (God)	Survival
4	Injury: E/ 660–662 Treatment: E/692–698	And Tlepolemus smote Sarpedon upon the left thigh with his long spear, and the point sped through furiously and grazed the bone; Then his goodly comrades made godlike Sarpedon to sit beneath a beauteous oak of Zeus that beareth the aegis, [695] and forth from his thigh valiant Pelagon, that was his dear comrade, thrust the spear of ash; and his spirit failed him, and down over his eyes a mist was shed. Howbeit he revived, and the breath of the North Wind as it blew upon him made him to live again after in grievous wise he had breathed forth his spirit.	Spear extraction and wind blowing (see Fig. 8.2)	Thigh injury from a spear	Sarpedon (Trojan ally)/Pelagon (Trojan)	Survival
5	E/794–798		Wound care and cleaning	Shoulder injury from an arrow	Diomed (Greek)/ Diomed (Greek)	Survival
6	Λ/396–397		Arrow extraction	Foot injury from an arrow	Diomed (Greek)/ Diomed (Greek)	Survival
7	Λ/455–458		Spear extraction	Thorax injury from a spear	Odysseus (Greek)/ Odysseus (Greek)	Survival
8	Λ/844–849		Arrow extraction and wound care, herbs for the pain	Thigh injury from an arrow	Eurypylus (Greek)/ Patroclus (Greek)	Survival

(Continued)

Table 8.3 (Continued)

Case number	Book/ verses for treatment	Text	Type of treatment	Type of trauma	Injured/care provider	Final outcome following treatment
9	N/598–599		Spear extraction and bandage cover of the wound	Hand injury from a spear	Helenus (Trojan)/ Agenor (Trojan)	Survival
10	Ξ/6–7		Wound care and cleaning, warm bath	Shoulder injury from an arrow	Machaon (Greek)/ Hecamede (Greek)	Survival
11	Ξ/424–439		Wound care and cleaning	Thorax injury from a stone	Hector (Trojan)/ Polydamas, Aeneas, Agenor, Sarpedon, Glaucus (Trojans)	Survival
12	Π/520–530		Pressure application to the trauma	Upper extremity arrow trauma	Glaucus (Trojan)/ Glaucus (Trojan)	Survival

Translations from Homer (1924).

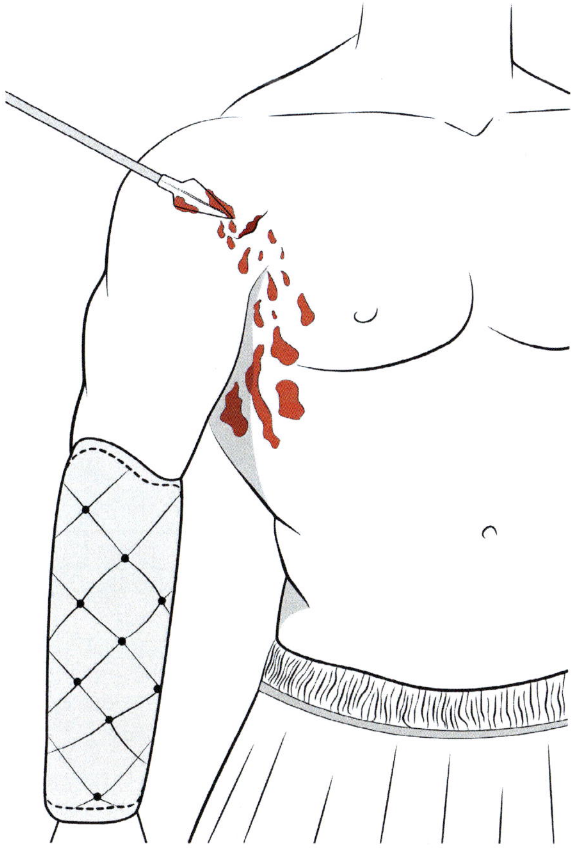

Figure 8.1 Diomedes' shoulder injury.
Shown is the arrow shot by Pandaros injuring Diomedes' shoulder which was removed by Sthenelus. See Case 2 in Table 8.3 for discussion.
Source: Image by Mileidy Fernandez after an image drawn by Anusha Lekshminarayanan.

His medical interventions spanned wound care, alleviation of pain, and measures to prevent hypothermia, including the provision of warm baths for trauma patients—an insight gleaned from various references. Homeric Greeks displayed a fundamental understanding of disinfection, emphasizing the importance of cleanliness. The "surgical" tools were prepped by cleaning them with seawater, as suggested in The *Iliad*, Rhapsody A, Verses 312–314, in conjunction with other sources (Passas, 1960).

Paramedics

Throughout the 19th and 20th centuries, paramedics emerged as pivotal figures in wartime trauma care, mirroring their crucial role in modern times. Acting as the

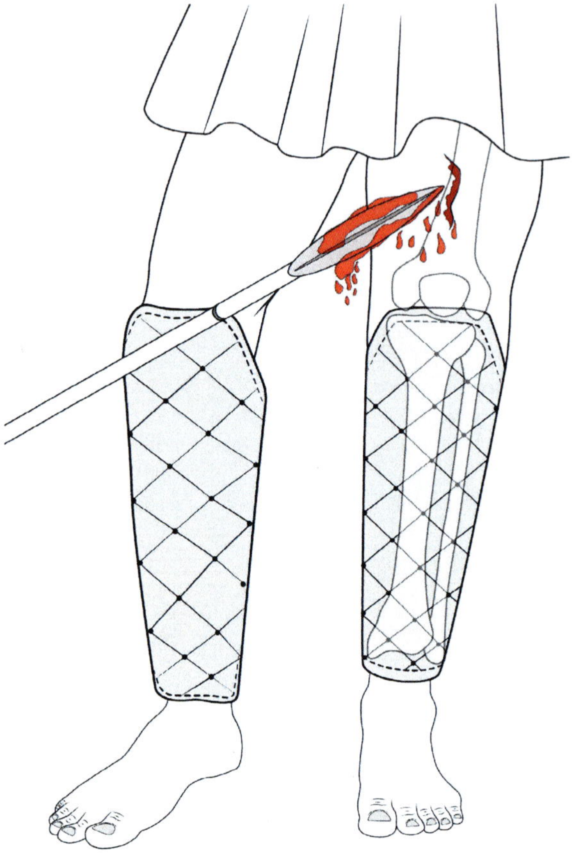

Figure 8.2 Sarpedon's left thigh injury.
Shown is the spear thrown by Tlepolemus injuring Sarpedon's thigh which was removed by Pelagon. See Case 4 in Table 8.3 for discussion.
Source: Image by Mileidy Fernandez after an image drawn by Anusha Lekshminarayanan.

initial responders to the injured, they administer primary medical aid and on-site pain relief. Their responsibilities extend to triage, enabling the prioritized transportation of wounded soldiers based on the severity of their injuries or the strategic necessities of battle and warfare. Serving as the vital connection between the battlefield and the essential medical facilities, paramedics form an integral link in the chain (Picton, 2014).

The poet dedicates significant attention to the role of paramedics, exemplified by figures like Mecisteus and Alastor, two Greek warriors embodying this crucial function (Koutserimpas & Samonis, 2017). Their duty included the transportation of injured comrades to Machaon's tent within the camp, ensuring urgent medical attention—a detail highlighted in The *Iliad* (Cases 3 and 5 of Table 8.1).

Triage

Homer also delves into the critical concept of triage, particularly significant in wartime scenarios. Triage within the Homeric Greek army followed a "strategic" approach, where priority often hinged not solely on the severity of trauma but on the injured warrior's role within the army (Koutserimpas & Samonis, 2017). An illustrative instance is the urgent transport of Machaon back to the Greek camp. Soldiers risked their lives, creating a human shield around him, exemplifying the prioritization of a "trauma specialist" surgeon injured in battle (The *Iliad*, Rhapsody Λ, Verses 510–515). Considering contemporary scenarios, where the importance of a military figure wounded on the battlefield prompts similar responses, it becomes evident that the principles of war triage have endured remarkably

Nursing

Homer portrays Hecamede as a battlefield nurse during the Trojan War, skilled in tending to the wounded and adept in treating bleeding injuries (Balanika & Baltas, 2014). As a token of appreciation for his wisdom, she had been bestowed upon Nestor, King of Pylos. Hecamede is depicted as 'beautiful and 'skilled as a goddess (Balanika & Baltas, 2014).

In the *Iliad*, specifically in Rhapsody Ξ, Hecamede is reported to attend to the physician Machaon who is injured in the right shoulder Case 10 in Table 8.3. (Did the warm bath help prevent hypothermia after blood loss?)

Another depiction of Hecamede in the *Iliad* shows her serving Nestor a medicinal drink called "Pramnian wine" or Kykeon, noted for its pharmaceutical properties (Fig. 8.3). Rhapsody Λ, lines 624–642, provides a detailed description of Hecamede's meticulous preparation of this drink.

Healthcare providers and trauma management

In the *Iliad*, Homer portrays Machaon and Podalirius as esteemed combat physicians, both sons of Asclepios (Fig. 8.4). Machaon, a surgeon, receives mention 11 times, while Podalirius, a physician, is referenced twice. Podalirius specialized in wound care, applying a unique pain-relieving ointment, while Machaon played a pivotal role: tending to King Menelaus' arrow injury (Case 1 in Table 8.3). (See also Fig. 8.3 in Apostolakis et al. (2010) for an artistic representation by F. Nenci (1837) of the arrow extraction.) Machaon later became a focal point himself, with Greeks rallying around him to safeguard against enemy capture or harm when he sustained an arrow wound to his right shoulder (Koutserimpas & Samonis, 2018).

Machaon's significance is underscored by Achilles' concern for his well-being, sending his closest friend Patroclus to assess Machaon's condition. This emphasizes the esteemed position of the physician, whose value equaled that of numerous men due to his life-saving knowledge and abilities (Λ, 611–615). While the brothers held chief medical roles, other warriors possessed trauma management expertise. Achilles

Figure 8.3 Hecamede serving Nestor kykeon. Image on attic-red figure cup shows Hecamede mixing kykeon for Nestor.
Source: Public Domain. https://en.m.wikipedia.org/wiki/File:Briseis_Phoinix_Louvre_G152.jpg.

Figure 8.4 Asklepios and his two doctor sons Podalerios and Machaon.
Relief in the form of a shrine: center Asklepios; behind him his two doctor sons Podalerios and Machaon—the two main physicians of the Greek army during the Trojan War—then his three daughters: Iaso, Akeso, and Panakeia. Opposite four worshippers.
Source: Data from "Wellcome Library, London."

Figure 8.5 Chiron the centaur teaching Achilles the medical arts.
Achilles learning from Chiron the centaur, mixing herbs and ointments for medical purposes. Etching by A.M. Zanetti, 1758, after G. Zompini after G.B. Castiglione.
Source: Data from "Wellcome Library, London." Copyrighted work available under CC-BY Attribution 4·0 International license.

Figure 8.6 Sthenelos taking care of a wound on the index finger of the fellow warrior Diomedes.
Sthenelos taking care of a wound at the index finger of the great warrior Diomedes. Diomedes was one of the most decorated Greek warriors. Ink drawing after a Chalcidian neck-amphora circa 550B.C. Ink drawing after a Chalcidian neck-amphora circa 550B.C. (edited in black and white).
Source: Credits to "Wellcome Library, London." Copyrighted work available under CC-BY Attribution 4·0 International license.

learned the medical arts from centaur Chiron and passed this knowledge to Patroclus, who surgically addressed a femoral injury by widening the incision to remove an arrow (Λ, 844–4=849) (Fig. 8.5). Nestor, king of Pylos, attended to Machaon's injury (Λ, 510–515). Additionally, Sthenelos attended to fellow warrior Diomed's injury (Fig. 8.6) and Diomed himself later tended to his own wound (E, 112). Sarpedon (Table 8.3, case 4) found relief from pain after exposure to the north wind's blowing, potentially linked to enhanced oxygen delivery (Balanika & Baltas, 2014; Koutserimpas et al., 2017). After spear extraction Helenus (Table 8.3, case 9) was provided with a woolen bandage for his injury.

Even the gods had their designated physician, Paeeon, who was summoned as necessary.

The mortality rate resulting from treated trauma cases was 5% (Koutserimpas et al., 2017) but this is likely skewed as more severe injuries typically resulted in rapid death usually from laceration of major vessels (see Chapters 9–13).

Conclusions

Among the thousands of verses in the *Iliad*, there is valuable information about the medical practice of that time. Although this epic poem is not considered a medical document, insights about the healthcare structure during war, as well as trauma management are provided. The basic principles of triage, the role of nurses, paramedics, and the esteemed role of the physician are documented. These principles have been applied throughout the years in modern wartimes and trauma management. It is of note that the description of these medical details, especially regarding the management of traumas, has led to a belief that Homer himself could also have been a surgeon or a physician. His depictions of these wounds and their treatment require some sort of medical and/or anatomy. The study of Homer's *Iliad*, undoubtedly, provides valuable insights about the medical practices and the principles of trauma treatment of the time.

References

Apostolakis, E., Apostolaki, G., Apostolaki, M., & Chorti, M. (2010). The reported thoracic injuries in Homer's *Iliad*. *Journal of Cardiothoracic Surgery, 5*. https://doi.org/10.1186/1749-8090-5-114.

Balanika, A. P., & Baltas, C. S. (2014). Hecamede: Homeric nurse of the battle-wounded in the Trojan War. *Journal of Medical Biography, 22*(1), 32–34. https://doi.org/10.1177/0967772013479261, http://www.jmb.rsmjournals.com.

Godquin, B. (1990). Was Homer a surgeon? *Chirurgie - Memoires de l'Academie de Chirurgie, 116*(2), 136–143.

Kakridis, I. (1986). *Greek mythology* (1st ed.).

Koutserimpas, C., & Samonis, G. (2017). Letter to the editor: Medical structures during war in the Homeric era: paramedics and 'expedition' hospitals. *Journal of Wound Care, 26*(3), 152–153. https://doi.org/10.12968/jowc.2017.26.3.152b.

Koutserimpas, C., & Samonis, G. (2018). Machaon: The first trauma surgeon in Western history? *Journal of Wound Care, 27*(10), 659–661. https://doi.org/10.12968/jowc.2018.27.10.659.

Koutserimpas, C., Alpantaki, K., & Samonis, G. (2017). Trauma management in Homer's *Iliad*. *International Wound Journal, 14*(4), 682–684. https://doi.org/10.1111/iwj.12672.

Passas, J. (1960). New encyclopedic dictionary.

Picton, C. (2014). Role of paramedics. *Emergency Nurse, 21*(10), 5. https://doi.org/10.7748/en2014.03.21.10.5.s1.

Upper extremity injuries in Homer

9

Tyler Lucas[1], *Anusha Lekshminarayanan*[2,3], *and Eric L. Altschuler*[4,5]
[1]Division of Orthopaedic Surgery, Department of Surgery, Metropolitan Hospital, New York, NY, United States, [2]Department of Physical Medicine and Rehabilitation, Westchester Medical Center, Valhalla, NY, United States, [3]Department of Rehabilitation Medicine, New York Medical College, Valhalla, NY, United States, [4]Essen Health Care, Bronx, NY, United States, [5]Kessler Institute for Rehabilitation, West Orange, NJ, United States

Upper limb injuries in the *Iliad* are listed in Table 9.1 and entry points for weapons are shown in Fig. 9.1. (See also (Apostolakis et al., 2010; Kömürcü et al., 2014; Saunders, 1999; Thompson, 1952).) Weapons used to inflict upper extremity injuries in the *Iliad* (Table 9.1) and their fatality rate are listed in Table 9.2. (See also (Apostolakis et al., 2010; Kömürcü et al., 2014).)

Note 1

See Fig. 9.2. An injury to the axillary artery can cause exsanguination and rapid death if left untreated.

Fig. 9.3 demonstrates Case 8.

Fig. 9.4 shows the injury in Case 15.[1]

Fig. 9.5 demonstrates the injuries in Cases 17 and 18.

Note 2

A propos this case one of us (TL) recalls the following story: One evening when I was in middle school I was walking home with my friend after we finished playing ball outside. Out of nowhere—the assailant was never found—my friend was struck by an arrow in the front [anterior] of his shoulder. Luckily, the arrow had a straight (*not* a barbed) point and my friend quickly and easily pulled the arrow out. There was not much bleeding. Still, his mother took him to the emergency room where the wound was cleaned and bandaged up. The injury healed quickly and my friend suffered no lasting ill effects.

[1] Even for medical personnel this injury is not easy to look at. It is hard to believe that between treating the wounded and seeing the bodies of fallen warriors Homer himself did not have PTSD after the Trojan War. See Chapter 22 for more on this.

Medicine in Homer. DOI: https://doi.org/10.1016/B978-0-443-32978-4.00023-7

Table 9.1 Upper limb injuries listed in the *Iliad*.

Number	Citation	Injuries	Victim	Attacker	Weapon	Outcome	Brief analysis	Comments
1	2; 212–277	Shoulder	Thersites (G)	Odysseus (G)	Cane	survival	Soft tissue injury	See Chapter 4
2	5; 43–48	Shoulder and axillary artery	Phaistos (T)	Idomeneus (G)	Spear	Death	Vascular injury with bleeding	See Note 1 and Fig. 9.2
3	5; 76–83	Shoulder disarticulation/ forequarter amputation	Hypsemor (T)	Euryplos (G)	Sword	Death	Death from blood loss.	
4	5; 95–120	Shoulder	Diomedes (G)	Pandaros (T)	Arrow	Survival	Laceration	See Note 2 below and also Fig. 8.2 for treatment
5	5; 144–148	Shoulder, clavicular fracture, amputation	Hyperion (T)	Diomedes (G)	Spear	Death	Forequarter amputation. discussed by Thompson (1952). NA	
6	5; 330–430	Hand	Aphrodite (Goddess)	Diomedes (G)	Spear	Survival	Laceration	Injury possibly at the junction of the distal radius and scapholunate
7	5; 395, 404	Shoulder	Hades (God)	Herakles or Amphitryoniad-es (Zeus' son)	Arrow	Survival	Laceration	

8	5; 576–579	Clavicle fracture	Pylaimenes (T)	Menelaus (G)	Spear	Death	Pneumohemothorax?	Comminuted open fracture (also likely injured one of the great vessels: subclavian vein, innominate or brachiocephalic artery). See Fig. 9.3
9	7; 13–16	Shoulder	Iphinos (G)	Glaukos (T)	Spear	Death		
10	8; 320–329	Hand	Teucer (G)	Hector (T)	Stone	Survival	Soft tissue injury	Brachial plexus injury. See Chapter 15
11	11;418–422	Shoulder	Deioptis (T)	Odysseus (G)	Spear	Survival	Laceration	
12	11; 504–520	Shoulder	Machaon (G)	Aleksandros (T)	Arrow	Survival	Laceration	
13	12;329–391	Shoulder (not otherwise specified)	Glaukos (Lykian, Trojan ally)	Teucer (G)	Arrow	Survival	Laceration	
14	13; 516–539	Not specified	Deiphobos (T)	Meriones (G)	Spear	Survival	Laceration	
15	13; 581–600	Hand	Helenos (T)	Menelaus (G)	Spear	Survival	Laceration	"Spaghetti hand". See Fig. 9.4. For treatment of Helenos' injury, see Table 8.3, Case 9. The Trojan physician (Agenor) may

(Continued)

Table 9.1 (Continued)

Number	Citation	Injuries	Victim	Attacker	Weapon	Outcome	Brief analysis	Comments
								have provided more extensive treatment than Homer was willing to admit
16	14; 449–474	Shoulder	Prothoenor (G)	Polydamas (T)	Spear	Death		
17	15; 341–342	Shoulder	Deiokhos (G)	Aleksandros (T)	Spear	Death	Likely vascular and/or hemopneumo-thorax	See Fig. 9.5
18	15; 525–545	Shoulder	Dolops (T)	Menelaus (G)	Spear	Death	Likely vascular and/or hemopneumo-thorax	See Fig. 9.5
19	16; 284–293	Right shoulder	Pyraikhmes (T)	Patroclus (G)	Spear	Death		
20	16; 319–329	Shoulder and proximal humerus fracture	Maris (T)	Thrasymedes (G)	Spear	Death	Muscles penetrated would have been subscapularis or other rotator cuff muscles depending on entry and axillary artery.	See Fig. 9.6 and Note 3

21	16; 341–344	Right shoulder	Akamas (T)	Meriones (G)	Spear	Death	Laceration	
22	16; 788–806	Shoulder	Patroclus (G)	Phoebos Apollon	Hand	Survival	Contusion	
23	16; 806–816	Shoulder	Patroclus (G)	Eubhorbos (T)	Spear	Survival	Laceration	
24	17; 304–311	Clavicular fracture and subclavian artery injury	Schedios (G)	Hector (T)	Spear	Death		
25	17; 593–600	Shoulder open fracture	Peneleos (G)	Polydamas (T)	Spear	Survival	Lacerations	"Cut clear to the bone"
26	20; 478–480	Elbow open fracture "at the point where the sinews of the elbow join"	Deucalion (T)	Achilles (G)	Spear	Survival (but see comme-nt)	sinews likely biceps tendon	Deucalion was immediately then killed (lines 481–483) by a sword blow to the neck. See Case 46 in Table 13.2
27	21; 116–120	Clavicular area	Lycaon (T)	Achilles (G)	Sward	Death		
28	21; 161–167	Elbow	Achilles (G)	Asteropaios (T)	Spear	Survival	Grazing	
29	22; 306–404	Subclavicular area and suprasternal notch	Hector (T)	Achilles (G)	Sword	Death		
30	23; 262–652	Elbow injury	Eumelos (G)		Chariot race and broken yoke	Survival	Abrasion and bruises	Eumelos also sustained abrasions to the eyebrow/ forehead area [frontal bone] and checks [zygoma]

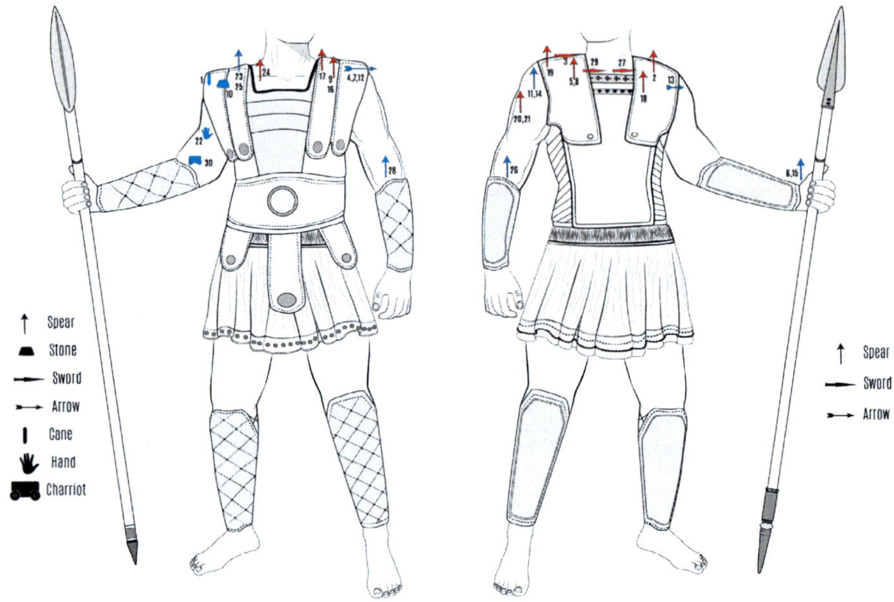

(A) (B)

Figure 9.1 Locator diagrams for upper extremity injuries. Upper extremity locator figure demonstrating injuries of Greek (G) (A) and Trojan (T) (B) victims. Red color indicates mortality, and blue color indicates survival. Case numbers as per Table 9.1 are also indicated in the figures.
Source: Image by Mileidy Fernandez after an image drawn by Anusha Lekshminarayanan.

Table 9.2 Weapons used to inflict upper limb injuries.

Weapon number	Weapon	Number of injuries	Number fatal (%)
1	Spear	19	11 (57.9%)
2	Arrow	4	0 (0%)
3	Sword	3	3 (100%)
4	Cane	1	0 (0%)
5	Chariot, yoke	1	0 (0%)
6	Hand	1	0 (0%)
7	Stone	1	0 (0%)
Total		30	14 (46.67%)

Note 3

For Case 20 (*Iliad*, Book 16, 321–325) "the point of the spear **shore** the base of the arm away from the muscles, and utterly brake asunder the bone," both Thompson

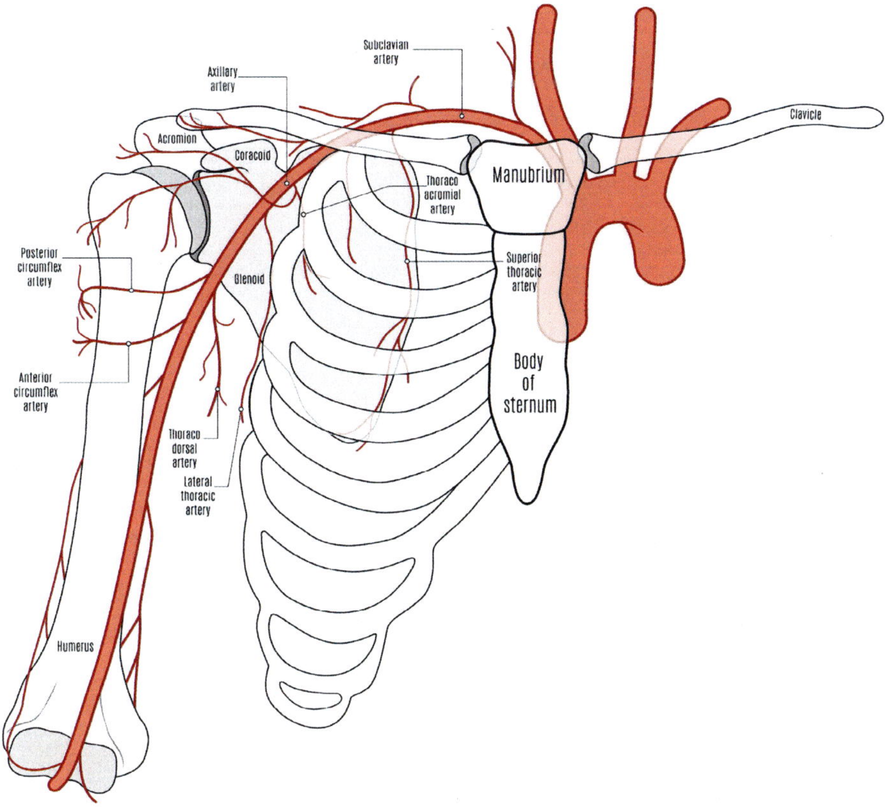

Figure 9.2 Axillary artery and its branches.
Source: Image by Mileidy Fernandez after an image drawn by Anusha Lekshminarayanan.

(Thompson, 1952) (a surgeon) and Saunders (Saunders, 1999) (a physician and classical scholar), find a weapon/injury mismatch noting that Homer does not use a spear to cause a shear injury.

Iliad, Book 16, lines 321–325 (Greek and translation from (Homer, 1924). This work is licensed under a Creative Commons Attribution-ShareAlike 3.0 United States License.)

στὰς πρόσθεν νέκυος: τοῦ δ᾽ ἀντίθεος Θρασυμήδης
ἔφθη ὀρεξάμενος πρὶν οὐτάσαι, οὐδ᾽ ἀφάμαρτεν,
ὦμον ἄφαρ: πρυμνὸν δὲ βραχίονα δουρὸς ἀκωκὴ
δρύψ᾽ ἀπὸ μυώνων, ἀπὸ δ᾽ ὀστέον ἄχρις ἄραξε:
δούπησεν δὲ πεσών, κατὰ δὲ σκότος ὄσσε κάλυψεν.

But Maris, hard at hand, [320] rushed upon Antilochus with his spear, wroth for his brother's sake, and took his stand before the dead; howbeit godlike Thrasymedes was too quick for him, and forthwith ere his foe could thrust, smote upon his shoulder, and

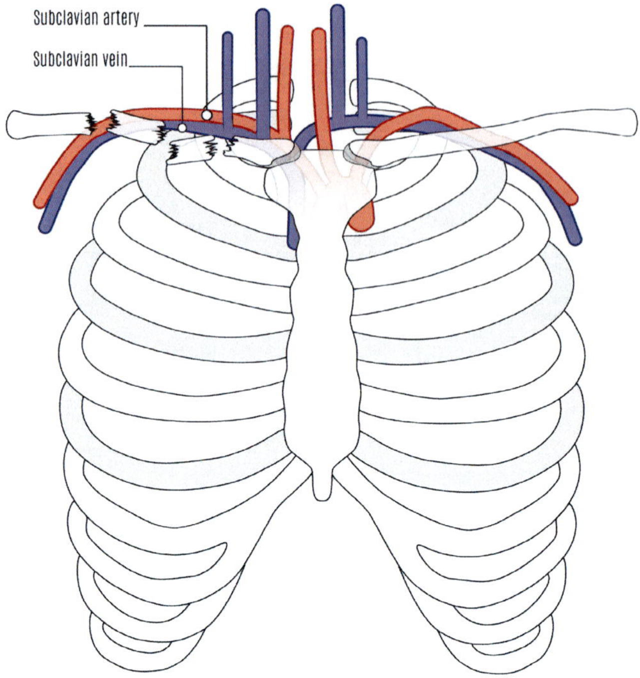

Figure 9.3 Open communited clavicle fracture. An open, comminuted clavicle fracture can result in injury of the subclavian artery, brachiocephalic artery, or the innominate vein (Case 8). *Source*: Image by Mileidy Fernandez after an image drawn by Anusha Lekshminarayanan.

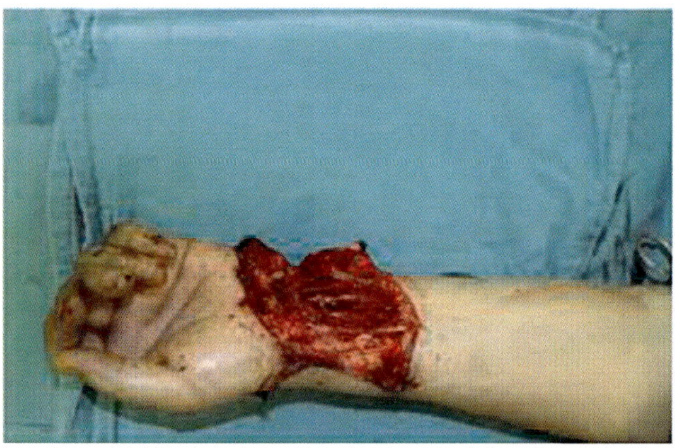

Figure 9.4 "Spaghetti hand." Modern day example of what is knowm today as "spaghetti hand." *Source*: Boynuyogun, D. M., et al. (2021). Combined nerve, vessel, and tendon injuries of the volar wrist: Multidisciplinary treatment and functional outcomes [Lésions combinées des nerfs, vaisseaux et tendons de la face palmaire du poignet: Traitement multidisciplinaire et résultats fonctionnels]. *Hand Surgery and Rehabilitation*, 40 (6), 729–736. Elsevier Masson SAS. Used with permission.

Figure 9.5 Spear entering under scapula. Drawing of a spear entering under the scapula and coming out through the chest. Pink color indicates the lung and red color shown around the right lung indicates hemothorax.
Source: Image by Mileidy Fernandez after an image drawn by Anusha Lekshminarayanan.

missed not; but the point of the spear **shore** the base of the arm away from the muscles, and utterly brake asunder the bone; [325] and he fell with a thud, and darkness enfolded his eyes.

For Homer δρύπτω does means shearing. For example, in the *Odyssey* Book 2, line 153 (as pointed out by (Saunders, 1999)) eagles **tear** into each other with their talons:

Odyssey, Book 2, line 153 (Greek and translation from (Homer, 1919). This work is licensed under a Creative Commons Attribution-ShareAlike 3.0 United States License.)

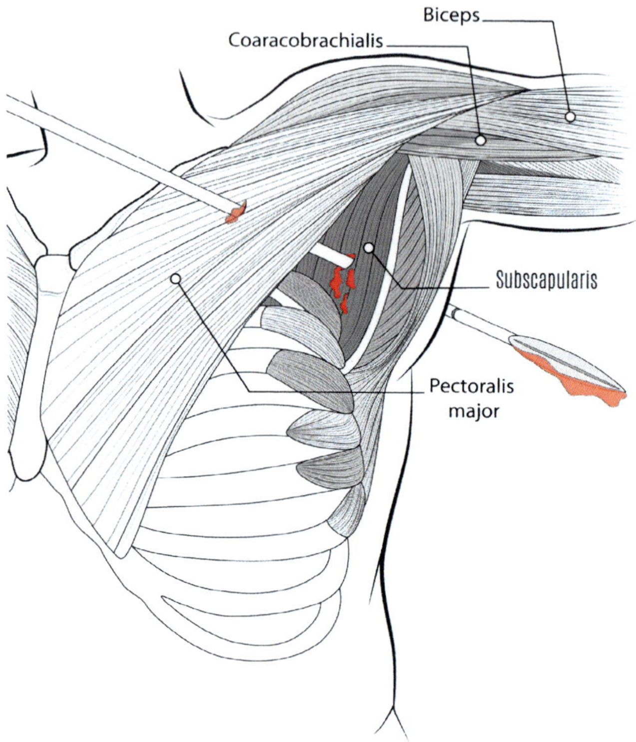

Figure 9.6 Maris' spearing. Muscles are possibly torn in Maris' spearing. See text for discussion.
Source: Figure by Mileidy Fernandez.

δρυψαμένω δ' ὀνύχεσσι παρειὰς ἀμφί τε δειρὰς
Then they [eagles] **tore** with their talons one another's cheeks and necks on either side

We think that the injury/weapon discrepancy may be able to be resolved with the spear not shearing the muscles themselves but the muscles from the bone. In this case, the muscles affected by the anterior blow of the spear would be the pectoralis major and subscapularis muscles (see Fig. 9.6). (The long head of the biceps femoris and the coracobrachialis also could have been injured).

The fatality rate in the *Iliad* for injuries to the upper extremity (Table 9.2) is far lower than for injuries to other body areas (see Tables 10.2, 11.2, 12.2 and 13.3). However, as with vascular injuries in other parts of the body, laceration of the axillary artery or other major vessels in the chest in Homer (see Fig. 9.2, Fig. 9.3, Fig. 9.5 and Cases 2, 3, 8, 17, 18, 24, 27, and 29 in Table 9.1) is invariably lethal.

References

Apostolakis, E., Apostolaki, G., Apostolaki, M., & Chorti, M. (2010). The reported thoracic injuries in Homer's *Iliad*. *Journal of Cardiothoracic Surgery, 5*. https://doi.org/10.1186/1749-8090-5-114.

Homer (1919). *The Odyssey*. Harvard University Press.

Homer (1924). *The Iliad*. Harvard University Press.

Kömürcü, E., Tok, F., Simşek, A., & Ozçakar, L. (2014). Musculoskeletal injuries in Homer's *Iliad*: The War of Troy revisited. *American Journal of Physical Medicine & Rehabilitation, 93*(4), 335–341. https://doi.org/10.1097/PHM.0000000000000051.

Saunders, K. B. (1999). The wounds in "*Iliad*" 13-16. *Classical Quarterly, 49*(2), 345–363.

Thompson, A. R. (1952). Homer as a surgical anatomist. *Proceedings of the Royal Society of Medicine, 45*(11), 765–767.

Thoracic trauma in Homer

10

Anusha Lekshminarayanan[1,2], *Tyler Lucas*[3], *and Eric L. Altschuler*[4,5]
[1]Department of Physical Medicine and Rehabilitation, Westchester Medical Center, Valhalla, NY, United States, [2]Department of Rehabilitation Medicine, New York Medical College, Valhalla, NY, United States, [3]Division of Orthopaedic Surgery, Department of Surgery, Metropolitan Hospital, New York, NY, United States, [4]Essen Health Care, Bronx, NY, United States, [5]Kessler Institute for Rehabilitation, West Orange, NJ, United States

The article by Apostolakis et al. (2010) has an excellent appendix that lists and catalogs the thoracic injuries in the *Iliad*. Following this appendix, the injuries in Books 1–12 of the *Iliad* are listed in Table 10.1 and the injuries in Books 13–24 are listed in Table 10.2. Apostolakis et al. (2010) take an inclusive view of thoracic injuries so and include some upper extremity injuries (e.g., shoulder injuries) in the *Iliad* which are Chapter 9. The article by Santos (2000) discusses injuries to the chest in the *Iliad* and the article by Kömürcü et al. (2014) is also helpful. The locations of entry points for thoracic injuries are shown in Fig. 10.1 (corresponding to Table 10.1) and Fig. 10.2 (corresponding to Table 10.2).

A summary of weapons used and their lethality of the weapons is listed in Table 10.3. Thoracic injuries caused by a spear or sword are highly lethal.

Note 1

Iliad Book 13, lines 434–444:

This Alcathous did Poseidon subdue beneath Idomeneus, [435] for he cast a spell upon his bright eyes and ensnared his glorious limbs that he might nowise flee backward nor avoid the spear; but as he stood fixed, even as a pillar or a tree, high and leafy, the warrior Idomeneus smote him with a thrust of his spear full upon the breast, [440] and clave his coat of bronze round about him, that aforetime ever warded death from his body, but now it rang harshly as it was cloven about the spear. And he fell with a thud, **and the spear was fixed in his heart, that still beating made the butt thereof to quiver**; howbeit, there at length did mighty Ares stay its fury.

(Translations from (Homer, 1924) This work is licensed under a Creative Commons Attribution-ShareAlike 3.0 United States License.)

It thus seems Idomeneus' spear lodged in Alcathous' left ventricle the contractions of which caused the spear to quiver (Saunders, 1999; Thompson, 1952).

Medicine in Homer. DOI: https://doi.org/10.1016/B978-0-443-32978-4.00024-9

Table 10.1 Thoracic injuries in the *Iliad* Books 1–12.

Number	Citation	Injuries	Victim	Attacker	Weapon	Outcome	Brief analysis	Comments
1	2; 265–270	Abrasion and pustule in the chest	Thersites (G)	Odysseus (G)	Scepter	Survival		See Chapter 4
2	4; 467–469	Open chest stab wound	Elephinor (G)	Aginor (T)	Spear	Death		
3	4; 473–488	Right mammary chest entry wound and exit through shoulder	Simoeisios (T)	Aias (G)	Spear	Death		
4	4; 527–531	Inframammary chest stab wound (first injury), abdomen (second, fatal injury)	Peiroos (T)	Thoas (G)	Spear (first wound) (Sword to abdom-en second wound)	Death		After wounding Peiroos in the chest with his spear and then removing the spear Thoas struck Peiroos in the abdomen with a sword killing him

Table 10.1 (Continued)

Number	Citation	Injuries	Victim	Attacker	Weapon	Outcome	Brief analysis	Comments
5	5;17–24	Sternal wound	Phegeus (T)	Diomedes (G)	Spear	Survival (fainting)		
6	5; 38–42	Interscapular area	Odeon (T)	Agamemnon (G)	Spear	Death	Enters in the back and exits through the ribs.	
7	5; 43–47	Right shoulder	Phaistos (T)	Idomeneus (G)	Spear	Death	Axillary artery laceratio-n.	See Case 2 in Table 9.1 and Fig. 9.2
8	5; 55–58	Interscapular area	Scamandrus (T)	Menelaus (G)	Spear	Death	Entry in the back with injury quite possibly to the heart and spinal cord.	
9	5; 79–83	Right shoulder/ arm disarticu-lation	Ipsinor (T)	Eurypylos (G)	Sword	Death		See Case 3, Table 9.1

(Continued)

Table 10.1 (Continued)

Number	Citation	Injuries	Victim	Attacker	Weapon	Outcome	Brief analysis	Comments
10	5; 95–100	Right shoulder, chest stab wound	Diomedes (G)	Pandarus (T)	Arrow	Survival		
11, 12	5; 144–148	Infraclavicular chest stab wound Forequarter amputation.	Astynoos (T) Hyperrion (T)	Diomedes (G) Diomedes (G)	Spear Spear	Death Death		See Case 4, Table 9.1
13	5; 280–282	Chest stab wound	Diomedes (G)	Pandarus (T)	Spear	Survival		See Case 5, Table 9.1
14	5; 392–395	Right mammary gland stab	Hera (Goddess)	Amphitryoniades (Zeus' son)	Arrow	Survival		Healed injury
15	5; 395–400	Shoulder	Hades (God)	Amphitryoniades (Zeus' son)	Arrow	Survival		Healed injury
16	5; 578–579	Clavicle	Pylaimenes (T)	Menelaus (G)	Spear	Death		See Case 8, Table 9.1, and Fig. 9.3
17	7; 13–16	Shoulder	Iphinoos (G)	Glaukos (T)	Spear	Death		
18	7; 268–272	Chest and abdomen compression	Hector (T)	Aias (G)	Stone	Survival (concus-sion, dizziness)		Old healed injury
19	8; 118–122	Mammary area	Eniopeus (T)	Diomedes (G)	Spear	Death		Case 8 in Table 9.1

Table 10.1 (Continued)

Number	Citation	Injuries	Victim	Attacker	Weapon	Outcome	Brief analysis	Comments
20	8; 257–260	Interscapular area	Agelaos (T)	Diomedes (G)	Spear	Death		Case 9 in Table 9.1
21	8; 300–308	Chest stab wound	Gorgythion (T)	Teukros (G)	Arrow	Death		
22	8; 309–315	Mammary area	Archeptolemos (T)	Teukros (G)	Arrow	Death		
23	8; 320–329	Clavicle, chest and neck injury	Teukros (G)	Hector (T)	Stone	Survival		Brachial plexus injury. See discussion of the injury in Chapter 15
24	11; 107–108	Infraclavicular area	Isos (T)	Agamemnon (G)	Spear	Death		
25	11; 143–144	Chest	Peisandros (T)	Agamemnon (G)	Spear	Death		
26	11; 259–261	Middle arm, chest and head	Koon (T)	Agamemnon (G)	Spear, sword	Arm injury (spear), then death from decapitation (sword).		Case 15 and Table 13.1

(*Continued*)

Table 10.1 (Continued)

Number	Citation	Injuries	Victim	Attacker	Weapon	Outcome	Brief analysis	Comments
27	11; 434–438	Intercostal muscles	Odysseus (G)	Sokos (T)	Spear	Survival	Lateral filleting injury to the superficial chest muscles.	The description suggests that Homer must have seen this injury
28	11; 446–449	Interscapular area	Sokos (T)	Odysseus (G)	Spear	Death		
29	11; 505–507	Shoulder	Mahaon (G)	Paris (T)	Arrow	Survival		Case 12 and Table 9.1

Source: Adapted from the Supplementary File from Apostolakis et al. (2010).

Table 10.2 Thoracic injuries in the *Iliad* Books 13–24.

Number	Citation	Injuries	Victim	Attacker	Weapon	Outcome	Brief analysis	Comments
30	13; 183–187	Chest stab wound	Amphimachos (G)	Hector (T)	Spear	Death		
31	13; 434–440	Sternum	Alkathoos (T)	Idomeneus (G)	Spear	Death	See Note 1.	Homer must have seen this injury—or gotten an immediate report—to describe it in such detail
32	13; 516–520	Shoulder and chest	Askalaphos (G)	Deiphobos (T)	Spear	Death		
33	14; 410–420	Chest and neck	Hector (T)	Aias (G)	Stone	Survival		See Note 2 for discussion
34	14; 449–452	Right shoulder	Prothoenor (G)	Polydamas (T)	Spear	Death		Case 16 and Table 9.1
35	15; 419–423	Chest stab wound	Kaletor (T)	Aias (G)	Spear	Death		
36	15; 520–524	Parasternal stab wound	Kroismos (T)	Meges (G)	Spear	Death		
37	15; 541–543	Posterior entry in shoulder area then through chest	Dolops (T)	Menelaus (G)	Spear	Death		
38	15; 575–584	Mammary area	Melanippos (T)	Antilochos (G)	Arrow	Death		
39	15; 645–652	Parasternal stab	Periphetes (G)	Hector (T)	Spear	Death		
40	16; 311–312	Parasternal stab	Thoas (T)	Menelaus (G)	Lance and/or spear	Death		

(*Continued*)

Table 10.2 (Continued)

Number	Citation	Injuries	Victim	Attacker	Weapon	Outcome	Brief analysis	Comments
41	16; 319–325	Head of humerus, shoulder	Maris (T)	Thrasymedes (G)	Spear	Death		See Case 20, Table 9.1, and Fig. 9.6
42	16; 342–344	Shoulder	Akamas (T)	Meriones (G)	Spear	Death		Case 21 and Table 9.1
43	16; 394–401	Sternum wound	Pronoos (T)	Patroclus (G)	Spear	Death		
44	16; 477–483	Precordium	Sarpedon (T)	Patroclus (G)	Spear	Death		See Note 3
45	16; 593–601	Sternal wound	Bathycles (G)	Glaucos (T)	Spear	Death		
46	16; 791–809	Interscapular area	Patroclus (A)	Euphorbos (T)	Javelin	Survival		
47	17; 304–311	Clavicle	Schedios (G)	Hector (T)	Javelin	Death		Case 24 and Table 9.1
48	17; 593–600	Shoulder	Peneleos (G)	Polydamas (T)	Spear	Survival		Case 25 and Table 9.1
49	20; 401–406	Interscapular area	Areithoos (T)	Achilles (G)	Spear	Death		
50	20; 487–489	Back (interscap-ular area)	Areithoos (T)	Achilles (G)	Spear	Death		
51	21: 116–120	Clavicle	Mykaon (T)	Achilles (G)	Sword	Death		Case 27 and Table 9.1
52	21; 403–408	Interscapular, neck area	Ares (God)	Athene (Goddess)	Stone	Survival		
53	21; 423–425	Breast	Aphrodite (Goddess)	Athene (Goddess)	Hand	Survival		

Source: Adapted from the Supplementary File from Apostolakis et al. (2010).

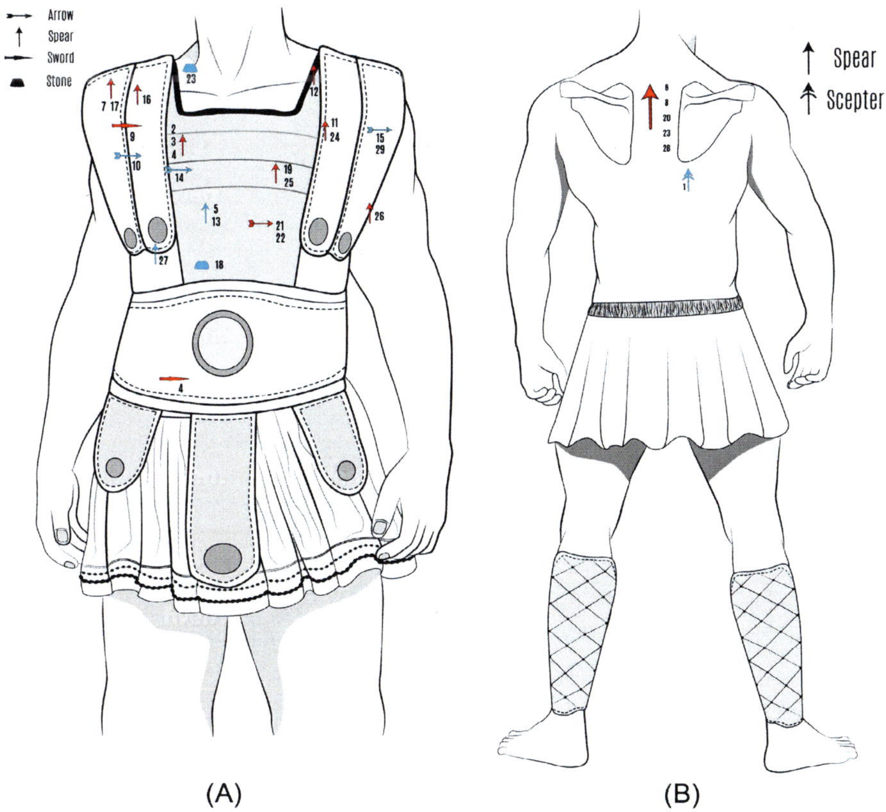

Figure 10.1 Locator Books 1–12. Entry points and weapons used to cause thoracic injuries mentioned in Books 1–12 of the *Iliad* are shown. Red indicates a fatal injury, and blue survival. The case number of the injury corresponding to the list in Table 10.1 is also given. Anterior entry wounds are shown in 10.1A and posterior entry wounds in 10.1.B.
Source: Image by Mileidy Fernandez after an image drawn by Anusha Lekshminarayanan.

Note 2

Hector's injury and sequela are described in Books 14 and 15 and his miraculous recovery—see below—later in Book 15:

Iliad Book 14, lines 409–439

But thereupon as he drew back, great Telamonian Aias smote him with a stone; [410] for many there were, props of the swift ships, that rolled amid their feet as they fought; of these he lifted one on high, and smote Hector on the **chest** over the shield-rim, hard by the neck, and set him whirling like a top with the blow; and he spun round and round. And even as when beneath the blast of father Zeus an oak falleth uprooted, [415] and a dread reek of brimstone ariseth therefrom—then verily courage no longer possesseth him that looketh thereon and standeth near by, for dread is the

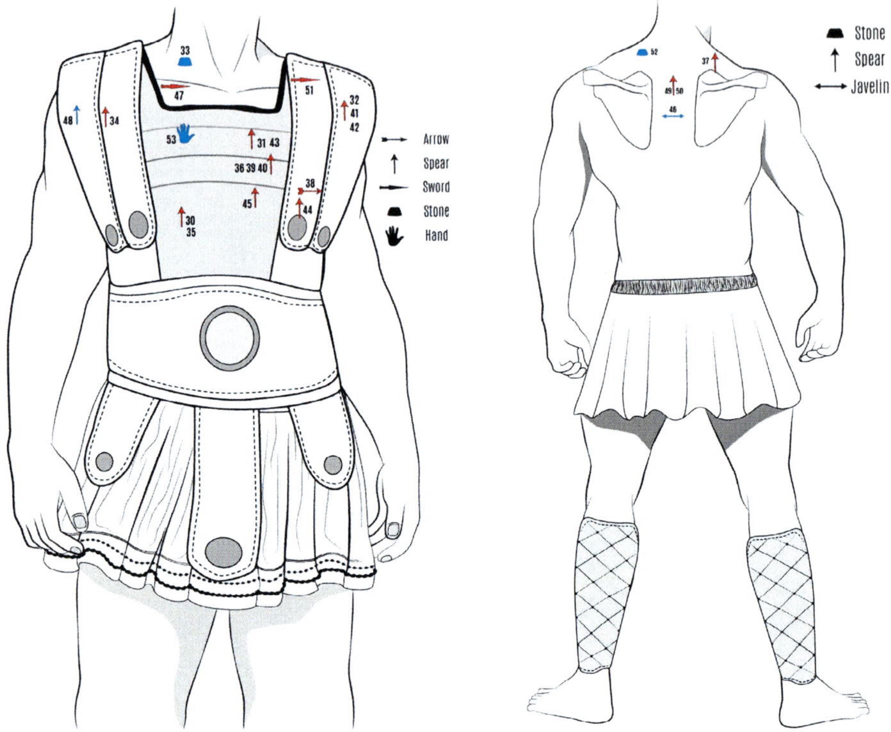

(A) (B)

Figure 10.2 Books 13–24 locator. Entry points and weapons used to cause thoracic injuries mentioned in Books 13–24 of the *Iliad* are shown. Red indicates a fatal injury, and blue survival. The case number of the injury corresponding to the list in Table 10.2 is also given. Anterior entry wounds are shown in 10.2A and posterior entry wounds in 10.2.B.
Image by Mileidy Fernandez after an image drawn by Anusha Lekshminarayanan.

bolt of great Zeus—even so fell mighty Hector forthwith to the ground in the dust. And the spear fell from his hand, but the shield was hurled upon him, [420] and the helm withal, and round about him rang his armor dight with bronze. Then with loud shouts they ran up, the sons of the Achaeans, hoping to drag him off, and they hurled their spears thick and fast; but no one availed to wound the shepherd of the host with thrust or with cast, for ere that might be, the bravest stood forth to guard him, [425] even Polydamas, and Aeneas, and goodly Agenor, and Sarpedon, leader of the Lycians, and peerless Glaucus withal, and of the rest was no man unheedful of him, but before him they held their round shields; and his comrades lifted him up in their arms and bare him forth from the toil of war until he came to the swift horses [430] that stood waiting for him at the rear of the battle and the conflict, with their charioteer and chariot richly dight. These bare him groaning heavily toward the city. But when they were now come to the ford of the fair-flowing river, even eddying

Table 10.3 Summary of weapons.

Weapon number	Weapon	Number of Injuries	Number fatal (%)
1	Spear	36	31 (86%)
2	Arrow	7	3 (43%)
3	Stone	4	0 (0%)
4	Sword	2	2 (100%)
5	Javelin	2	1 (50%)
5	Scepter	1	0 (0%)
7	Hand	1	0 (0%)
Total		53	37 (70%)

Source: Adapted from the Supplementary File from Apostolakis et al. (2010).

Xanthus, that immortal Zeus begat, [435] there they lifted him from the chariot to the ground and poured water upon him. And he revived, and looked up with his eyes, and kneeling on his knees **he vomited forth black blood**. Then again he sank back upon the ground, and both his eyes were enfolded in black night; and the blow still overwhelmed his spirit.

Saunders (1999) thinks that "vomited" in line 437 is better translated "coughed up" consistent with Hector having sustained a blow to the chest (line 412). That is, hemoptysis not hematemesis and we agree with this conclusion. We also agree with Saunders' analysis that after reviving (line 436) by kneeling (line 437, or sitting on his knees) he takes a body position that impedes venous return to the heart leading to fainting (line 439).

Iliad Book 15, lines 9–15

And Hector he [Zeus] saw lying on the plain, while about him sat his comrades, [10] and he was gasping with painful breath, distraught in mind, and vomiting blood; for not the weakest of the Achaeans was it that had smitten him. At sight of him the father of men and gods had pity, and with a dread glance from beneath his brows he spake to Hera, saying: "Hera", that art hard to deal with, it is the craft of thine evil wiles [15] that hath stayed goodly Hector from the fight, and hath driven the host in rout.

Iliad Book 15, lines 239–242

He [Apollo] found the son of wise-hearted Priam, even goodly Hector, [240] sitting up, for he lay no longer, and he was but newly gathering back his spirit, and knew his comrades round about him, and his gasping and his sweat had ceased, for the will of Zeus, that beareth the aegis, revived him.

Saunders (1999) notes that Hector's recovery in less than a day—rather than the week or more it should take—is miraculous and a device used for plot pacing. We do not disagree with this. Furthermore, we point out that in a "breaking the fourth wall" moment Homer makes us aware that he knows the recovery is miraculous when in Book 15, lines 286–288 Thoas says, "Now look you, verily a great marvel is this that

Figure 10.3 Euphronios krater. Panel from the Euphronios Krater depicting the death of Trojan ally Sarpedon. See text for further discussion of Sarpedon's injuries.
Source: By Jaime Ardiles-Arce (photographer). Krater by Euphronios (painter) and Euxitheos (potter). Wikipedia: contact us/photo submission, Public Domain, https://commons.wikimedia.org/w/index.php?curid=4944469

mine eyes behold, how that now he is risen again and hath avoided the fates, even Hector."

Note 3

Fig. 10.3 (also shown on the front cover) shows one panel from the amazing vase known as the Euphronios Krater (c.515 BCE) in honor of the artist. The panel depicts the injuries to staunch Trojan ally Sarpedon, son of Zeus from Lycia. As can be seen, Sarpedon is bleeding from the right thigh, right lower abdomen (injuries not mentioned in the *Iliad*), and the left chest (Case 44, Table 10.2)—his fatal wound. In a short interesting article (Johnson, 2001) pathologist Horton Johnson pointed out a small red dot in Sarpedon's left thigh which is also visible in the vase. This small wound in the left thigh which seems to be a healed wound was not commented on previously. Horton identifies this as the wound Sarpedon suffers in Book 5 (Case number 7 in Table 12.1, see also Fig. 8.2 and discussion thereof).[1] Horton concludes that Euphronios had a thorough knowledge of the *Iliad*. We agree.

[1] When Sarpedon was attacked by Tleplemos with a spear (Iliad Book 5, lines 627–698), he sustained a deep gaping thigh wound which must have exposed his quadriceps and femur bone. The spear was not removed immediately, likely avoiding exsanguination but increasing the chance of infection. Sarpedon survived but his treatment is not described in detail.

References

Apostolakis, E., Apostolaki, G., Apostolaki, M., & Chorti, M. (2010). The reported thoracic injuries in Homer's *Iliad*. *Journal of Cardiothoracic Surgery, 5*, 114. https://doi.org/10.1186/1749-8090-5-114.

Homer. (1924). *The Iliad*. Harvard University Press.

Johnson, H. (2001). The wounds of Sarpedon. *Lancet (London, England), 357*(9265).

Kömürcü, E., Tok, F., Simşek, A., & Ozçakar, L. (2014). Musculoskeletal injuries in Homer's *Iliad*: The War of Troy revisited. *American Journal of Physical Medicine & Rehabilitation, 93*(4), 335–341. https://doi.org/10.1097/PHM.0000000000000051.

Santos, G. H. (2000). Chest trauma during the battle of Troy: Ancient warfare and chest trauma. *The Annals of Thoracic Surgery, 69*(4), 1285–1287.

Saunders, K. B. (1999). The wounds in "*Iliad*" 13-16. *Classical Quarterly, 49*(2), 345–363.

Thompson, A. R. (1952). Homer as a surgical anatomist. *Proceedings of the Royal Society of Medicine, 45*(11), 765–767.

Abdominal trauma in Homer

11

Anusha Lekshminarayanan[1,2], *Tyler Lucas*[3], *and Eric L. Altschuler*[4,5]
[1]Department of Physical Medicine and Rehabilitation, Westchester Medical Center, Valhalla, NY, United States, [2]Department of Rehabilitation Medicine, New York Medical College, Valhalla, NY, United States, [3]Division of Orthopaedic Surgery, Department of Surgery, Metropolitan Hospital, New York, NY, United States, [4]Essen Health Care, Bronx, NY, United States, [5]Kessler Institute for Rehabilitation, West Orange, NJ, United States

Abdominal injuries are listed in Table 11.1. There are no publications specifically on abdominal injuries in the *Iliad*. Nomikos (2018) lists Book/line numbers for injuries and we have filled in the rest of the data in Table 11.1. In a few instances, for example, Cases 4, 10, 15, 16, and 32, the location of the injury in the abdomen is not specified. The locations for the rest of the injuries are shown in Figs. 11.1 and 11.2. The specificity of Homer's description of abdominal trauma is precise and impressive like his knowledge and interest in the anatomy of other body regions (see Chapters 9, 10, 12, and 13).[1] Data on weapons used and their lethality are given in Table 11.2.

Abdominal injuries are virtually always fatal. The only instances of survival are Menelaus (Case 1, Table 11.1), Ares who is a God (Case 8, Table 11.1), Case 11 where the abdominal injury was immediately followed by a fatal decapitation injury, and Cases 12, 16, and 21 in Table 11.1 for which the outcome may have been death but the result is not stated in the *Iliad*.

Note

When Menelaus is injured:

(Translations from Homer (1924). This work is licensed under a Creative Commons Attribution-ShareAlike 3.0 United States License.)

Iliad Book 4, lines 134–140

On the clasped belt lighted the bitter arrow, [135] and through the belt richly dight was it driven, and clean through the curiously wrought corselet did it force its way, and through the taslet which he wore, a screen for his flesh and a barrier against darts, wherein was his chiefest defense; yet even through this did it speed. So the arrow grazed the outermost flesh of the warrior [Menelaus], [140] and forthwith the dark blood flowed from the wound.

His brother Agamemnon—commander-in-chief—of the Greek expedition issued what we would call today a stat consult for surgeon Machaon of Tricca to assist:

[1] Given the trauma of war we would not expect 100% specificity of locations of abdominal injuries.

Medicine in Homer. DOI: https://doi.org/10.1016/B978-0-443-32978-4.00018-3

Table 11.1 Abdominal injuries in Homer's *Iliad*.

Case number	Citation	Injury	Victim	Attacker	Weapon	Outcome	Brief analysis	Comments
1	4; 130–147	Lower abdomen/ upper pelvis	Menelaus (G)	Pandaros (T)	Arrow	Survival		See Note and Fig. 11.3.
2	4; 489–493	Groin injury	Antiphus (T)	Leucus (G)	Spear	Death		See Case 2 in Table 12.1 for more on this injury
3	4; 525–527	Spear through navel	Diores (T)	Peiros (G)	Spear	Death		
4	4; 530–531	Abdomen	Thoas (Greek)	Piros (Trojan)	Sword	Death		
5	5; 66–68	Spear through buttock on through bladder	Phereclus (T)	Meriones (G)	Spear	Death		See also Case 4 inTable 12.1
6	5; 537–539	Lower belly	Deïcoön (T)	Agamemnon (G)	Spear	Death		
7	5; 611–617	Lower belly	Amphius (T)	Telamonian Aias (G)	Spear	Death		
8	5; 856–871	Lower most belly	Ares (god)	Diomedes (G)	Spear	Survival		
9	6; 62–65	Flank	Adrastus (T)	Agamemnon (G)	Spear	Death		
10	7;144–145	Middle	Ereuthalion (T)	Lykourgos (King of Thrace, Greece)	Spear	Death		Not in Trojan War (past event)
11	11; 260	(1) Beneath his shield, then immediately (2) decapitation	Coön (T)	Agamemnon (G)	Spear	Death from the decapitation injury		Also Case 15 in Table 13.1

12	11;424–425	Beneath his shield	Chersidamas (T)	Odysseus (G)	Spear	Death (apparently)		
13	11;577–579	Liver below the midriff	Apisaon (T)	Eurypylus (G)	Spear	Death		
14	12; 189	Girdle	Hippomachus (G)	Leonteus (T)	Spear	Death		
15	13; 370–372	Belly	Othryoneus (T)	Idomeneus (G)	Spear	Death		
16	13; 397–398	Belly	Asius (T)	Antilochus (G)	Spear	Not stated		
17	13; 410–412	Liver below the midriff	Hypsenor (T)	Deïphobus (G)	Spear	Death		
18	13; 506–508	Belly/bowels	Oenomaus (T)	Idomeneus (G)	Spear	Death		
19	13; 567–570	Between navel and groin	Adamas (T)	Meroniones (G)	Spear	Death		
20	13; 650–652	Buttock through urinary bladder	Harpalion (T)	Meriones (G)	Arrow	Death but not immediate		See also Case 12 in Table 12.1
21	14;446–447	Flank	Enops (T)	Aias (G)	Spear	Not stated		Renal artery injury
22	14;516–519	Flank/bowels	Hyperenor (T)	Agamemnon (G)	Spear	Death		
23	16; 317–319	Flank	Atymnius (T)	Antilochus (G)	Spear	Death		Renal artery laceration
24	16; 465	Lower belly	Thrasymelus (T)	Patroclus (G)	(Spear)	Death		
25	16; 820–821	Lowermost belly	Partroclus (G)	Hector (T)	Spear	Death		
26	17; 313–315	Belly/bowels	Phorcys (T)	Aias (G)	Spear	Death		
27	17; 347–349	Liver below the midriff	Apisaon (T)	Lycomedes (G)	Spear	Death		
28	17; 518–521	Lower belly	Aretus (T)	Automedon (G)	Spear	Death		
29	17; 578–579	Belt	Podes (T)	Menelaus (G)	Spear	Death		

(Continued)

Table 11.1 (Continued)

Case number	Citation	Injury	Victim	Attacker	Weapon	Outcome	Brief analysis	Comments
								Lower abdominal/ pelvis injury with death likely due to femoral artery laceration. (Case 15 in Table 12.1)
30	20; 413–420	Back through navel	Polydorus (T)	Achilles (G)	Spear	Death		
31	20; 469–471	Liver	Tros (T)	Achilles (G)	Sword	Death		"but he smote him upon the liver with his sword, and forth the liver slipped, and the dark blood welling forth therefrom filled his bosom"
32	20; 485–487	Belly	Rhigmus (T)	Achilles (G)	Spear	Death		
33	21; 180–182	Belly/navel	Asteropios (T)	Achilles (G)	Sword	Death		

Authors' own work.

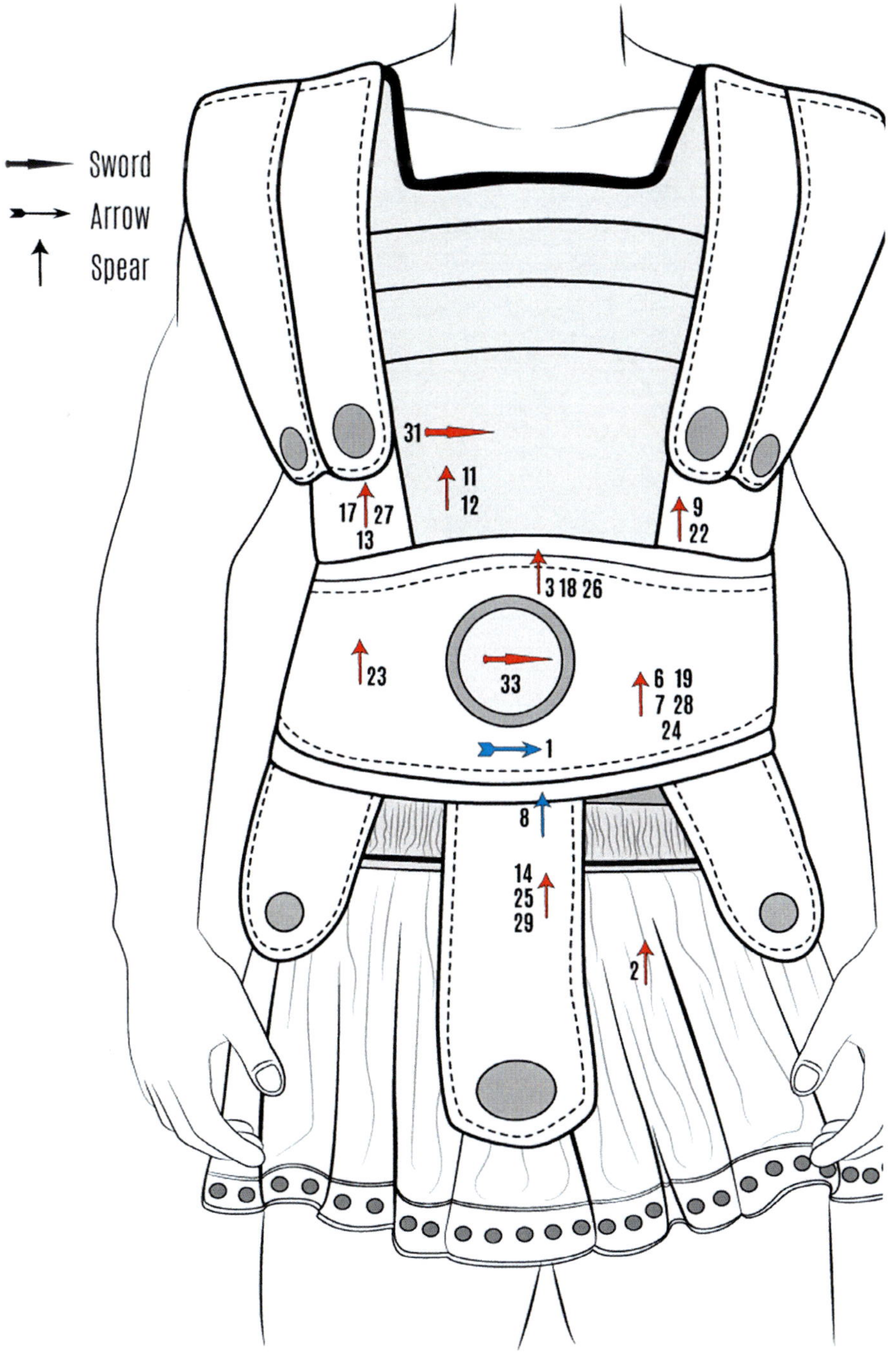

Figure 11.1 Anterior abdominal injuries. Location of anterior and lateral (flank) abdominal injuries. Injuries are indexed by case number in Table 11.1. For nonmidline (e.g., navel) entry points Homer does not indicate the side of the injury and in the figure a side has been chosen so as to maximize clarity of presentation. A red icon indicates a fatal injury, blue survival. *Source*: Image by Mileidy Fernandez after an image drawn by Anusha Lekshminarayanan.

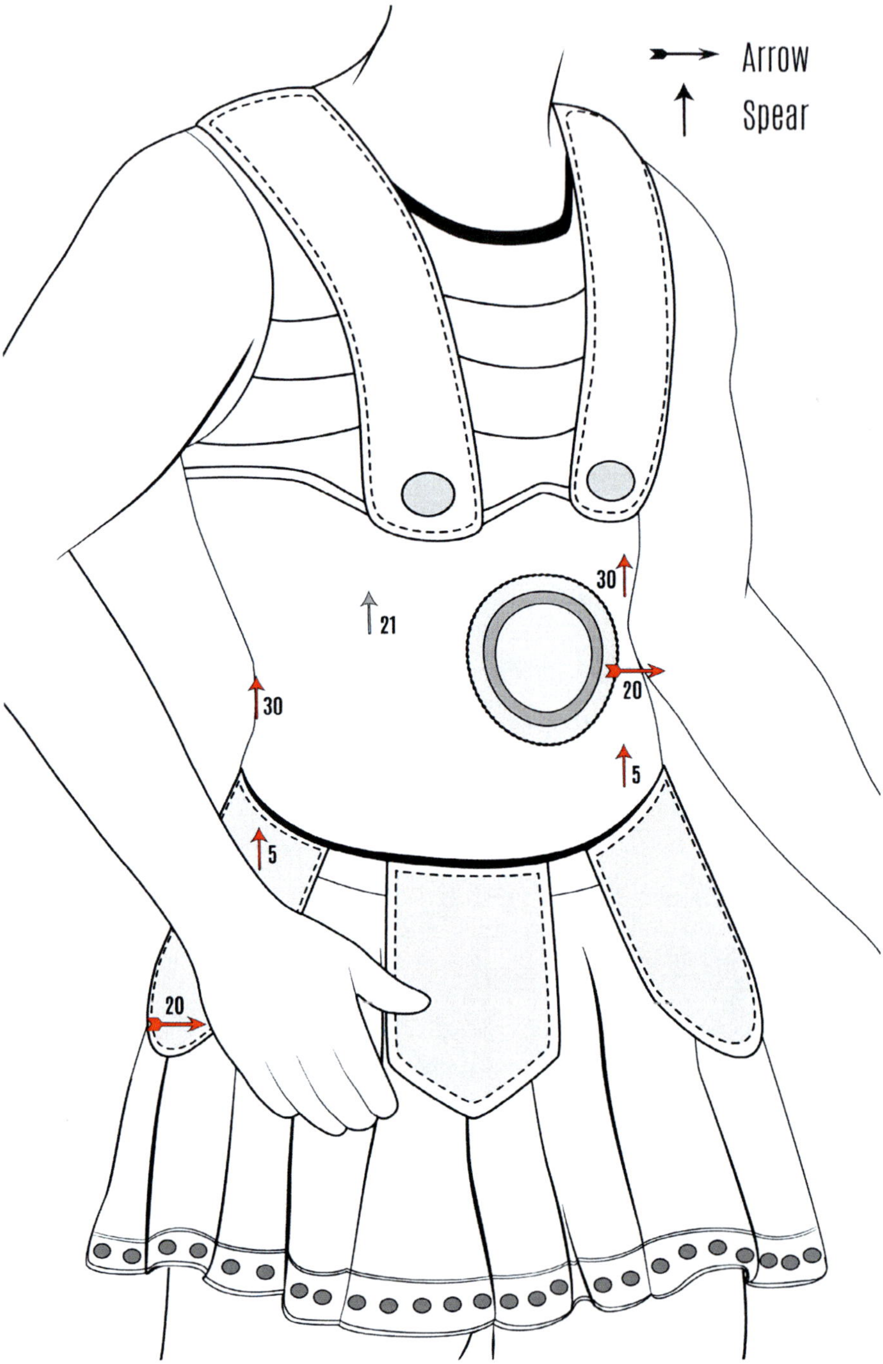

Figure 11.2 Posterior entry injuries. Location of abdominal injuries with posterior entry points. Injuries are indexed by case number in Table 11.1. A red icon indicates a fatal wound, and gray one for which the outcome was not stated.

Source: Image by Mileidy Fernandez after an image drawn by Anusha Lekshminarayanan.

Table 11.2 Summary of weapons.

Weapon number	Weapon	Number of injuries	Number fatal (%)
1	Spear	28	24 (86%)
2	Sword	3	3 (100%)
3	Arrow	2	1 (50%)
Total		33	28 (85%)

Authors' own work.

Iliad Book 4, lines 192–196

Therewith he spake to Talthybius, the godlike herald: "Talthybius, make haste to call hither Machaon, son of Asclepius, the peerless heeler, [195] to see warlike Menelaus, son of Atreus, whom some man well skilled in archery hath smitten with an arrow ... "

Machaon and his brother Podalirius were sons of Asclepius the fabled founder of Greek medicine. Machaon seems to have been the Chief of Surgery for the Greek expedition and Podalirius the Chief of Medicine. (See Chapters 3 and 8 for more on Machaon.) Homer describes Machaon's treatment of Menelaus (see also Fig. 11.3):

Iliad Book 4, line 212–219

The godlike hero [Machaon] came and stood in their midst, and straightway drew forth the arrow from the clasped belt; and as it was drawn forth the sharp barbs were broken backward. [215] And he loosed the flashing belt and the kilt beneath and the taslet that the coppersmiths fashioned. But when he saw the wound where the bitter arrow had lighted, he sucked out the blood, and with sure knowledge spread thereon soothing simples, which of old Cheiron had given to his father with kindly thought.

Where was Menelaus' injury? Just prior to Agamemnon's call for Machaon, Menelaus provides clues (*Iliad* Book 4, lines 183–191): But fair-haired Menelaus spake and heartened him [Agamemnon], saying: "Be thou of good cheer, neither affright in any wise the host of the Achaeans. [185] Not in a fatal spot hath the shaft been fixed; ere that my flashing belt stayed it, and the kilt beneath, and the taslet that the coppersmiths fashioned." Then in answer to him spake lord Agamemnon: "Would it may be so, dear Menelaus. [190] But the heeler shall search the wound and lay thereon simples that shall make thee cease from dark pains."

So Menelaus' belt and/or pelvic "cod piece" seem to have somewhat stopped/deflected the arrow. We think most likely Menelaus suffered a lower abdominal injury that did not pierce the transversalis fascia that covers the peritoneum (see Fig. 11.3)—or else he most surely would have died from peritonitis. Interestingly, the warriors were afraid to remove the arrow themselves—unlike when Sthenelus removed an arrow from Diomedes' shoulder (Case 2, Table 8.3)—but Machaon knew the arrow could be removed safely. He must have seen and treated this injury previously.

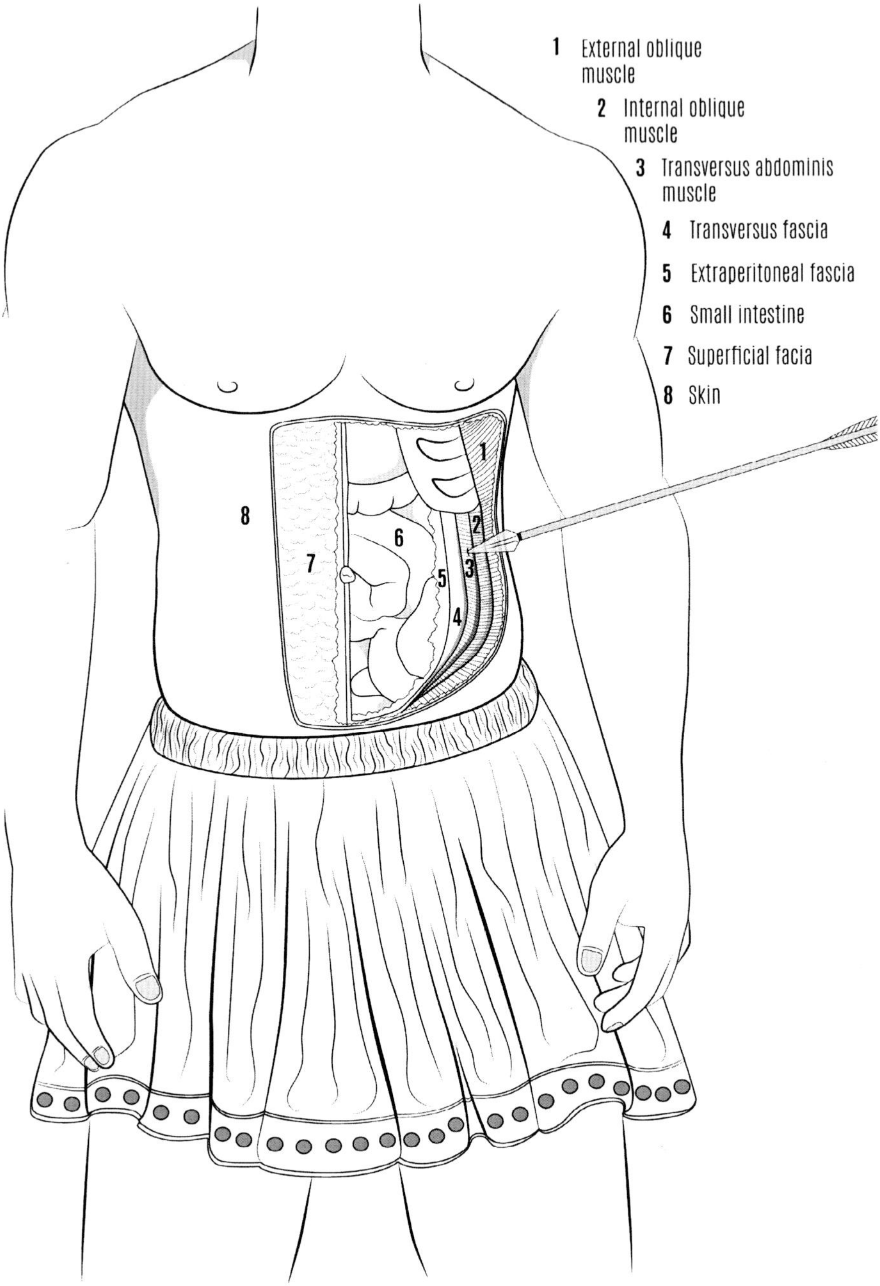

Figuer 11.3 Menelaus' injury. See also text for discussion.
Source: Image by Mileidy Fernandez after an image drawn by Anusha Lekshminarayanan.

References

Homer (1924). *The Iliad*. Harvard University Press.

Nomikos, N. (2018). Injuries in the Greek epics of Homer. *Chinese Journal of Traumatology=Zhonghua chuang shang za zhi, 21*(2), 109–112. https://doi.org/10.1016/j.cjtee.2017.09.005.

Lower extremity injuries in Homer

12

Tyler Lucas[1], *Anusha Lekshminarayanan*[2,3], *and Eric L. Altschuler*[4,5]
[1]Division of Orthopaedic Surgery, Department of Surgery, Metropolitan Hospital, New York, NY, United States, [2]Department of Physical Medicine and Rehabilitation, Westchester Medical Center, Valhalla, NY, United States, [3]Department of Rehabilitation Medicine, New York Medical College, Valhalla, NY, United States, [4]Essen Health Care, Bronx, NY, United States, [5]Kessler Institute for Rehabilitation, West Orange, NJ, United States

The pelvic and lower extremity injuries are listed in Table 12.1. The outcomes of the injuries and weapons used are listed in Table 12.2. The location of the injuries is illustrated diagrammatically in Fig. 12.1 and the entry points of injuries are shown in Fig. 12.2.

Note 1

Even in modern times whether due to injuries incurred in war (Pearl et al., 2003) or an accidental injury (Gitto et al., 2019) femoral artery lacerations (see Fig. 12.3) retain high morality and death can be rapid if the injury is left untreated (e.g., Gitto et al., 2019). Femoral artery lacerations, especially in the heat and midst of battle can be difficult to tourniquet. This may not even suffice to insure survival. For example, a femoral artery laceration to a Union soldier in the US Civil War which was deemed to be relatively minor despite rapid application of a tourniquet, transport away from the battle, convalescence at a major hospital center and an operation performed by leading surgeons ultimately were not enough to save the patient when the source of bleeding into a pseudoaneurysm could not be found (nor could it be located on autopsy either) (Heard, 1862). Without application of a tourniquet Confederate General Albert Sidney Johnston (1803–1862) became perhaps the most significant casualty in the extremely bloody Battle of Shiloh ("Battle of Shiloh," 2024)—one of the bloodiest of the US Civil War—dying rapidly following a popliteal artery (the main artery in the leg distal to the femoral artery) laceration. Thus Homer associates femoral artery lacerations with death.

Note 2

The anatomical location of the injury has been discussed and a number of possibilities have been suggested. We think the "two sinews" injured were the tibialis anterior

Medicine in Homer. DOI: https://doi.org/10.1016/B978-0-443-32978-4.00022-5

Table 12.1 Pelvis and lower extremity injuries in the *Iliad*.

Case number	Citation	Injury	Victim	Attacker	Weapon	Outcome	Brief analysis	Comments
1	2; 716–725	Soft tissue injury	Philoctetes (G)	Snake	Snake bite	Survival	–	Philoctetes' case featured in the works of authors after Homer (Johnson, 2003)
2	4; 489–493	Inguinal area pene-trating injury	Leukos (G)	Antiphos (T)	Spear	Death	Femoral artery laceration.	See Note 1 and Fig. 12.3
3	4; 517–538	Open ankle injury	Dicres (T)	Pyrus (G)	Stone (brok-en rock)	Death from intra-abdom-inal injury	Anterior tibialis and extensor digitorum longus before trifurcation laceration.	See Note 2 and Fig. 12.4
4	5; 59–68	Pelvis versus buttock	Phereklos (T)	Meriones (G)	Spear	Death	Spear entry in the buttock then penetrates the urinary bladder.	See Note 3 and Fig. 12.5

5	5; 297–317	Open femur acetabular joint fracture “broke both sinews”	Aeneas (T)	Diomedes (G)	Stone	Survival	Fractures but no vascular injury. Probably open femur/acetabular fracture and laceration of the rectus femoris and iliopsoas tendons (“both sinews”).	See Note 4 and Fig. 12.6
6	5; 528–540	Pelvis	Deikoon (T)	Agamemnon (G)	Spear	Death	Femoral artery laceration	see Fig. 12.3
7	5; 627–698	Thigh wound to bone	Sarpedon (G)	Tleplemos (T)	Spear	Survival	Femoral artery spared	For treatment, see Case 4 in Table 8.3 and discussion of Fig. 10.3
8	11; 338–375	Penetrating hip injury	Agastrophos (T)	Diomedes (G)	Spear	Death	Femoral artery laceration	See Note 1
9	11;375–381	Foot laceration	Diomedes (G)	Alexandros (T)	Arrow	Survival	See Fig. 12.7.	Diomedes laughs off injury (lines 385–395)
10	11; 575–592	Thigh injury	Euryplos (G)	Alexandros (T)	Arrow	Survival	Thigh injury, not otherwise specified (quadriceps injury)	
11	13; 560–575	Pelvis	Adamas (T)	Meriones (G)	Spear	Death	Femoral artery laceration	See Note 1

(*Continued*)

Table 12.1 (Continued)

Case number	Citation	Injury	Victim	Attacker	Weapon	Outcome	Brief analysis	Comments
12	13; 643–659	Hip and pelvis	Harpalion (T)	Meriones (G)	Arrow	Death	Entry in the buttock, then pierces the urinary bladder.	Death was not immediate. See Note 3 and Fig. 12.5
13	16; 306–311	Pelvic fracture and/or (?) hip injury	Areilykos (T)	Patroclus (G)	Spear	Death	Pelvic injury versus iliac crest injury	
14	16; 313–316	Hip and/or thigh injury	Amphiklos (T)	Meges (G)	Spear	Death	-	Sinews referred to by Homer could be rectus femoris tendon and/or hip capsule ligaments
15	17; 575–581	Pelvic pene-trating injury	Podes (T)	Menelaus (G)	Spear	Death	Femoral artery laceration	Hector commented that Podes was “good at the feast” (i.e., a drinking buddy)

16	20; 407–420	Pelvis	Polydoros (T)	Achilles (G)	Spear	Death	Achilles killed Polydoros by a spear from the lumbar paraspinal area. The spear exited anteriorly by the navel extruding his intestines—posterior to anterior injury. See Fig. 12.2.	The injury in Case 11 above was anterior to posterior

Authors' own work.

Table 12.2 Weapons used to inflict pelvis and lower extremity injuries in the *Iliad*.

Weapon number	Weapon used	Number of injuries	Number fatal (%)
1	Spear	10	9 (90%)
2	Arrow	3	1 (33%)
3	Stone	2	1 (after subsequent spearing) (50%)
4	Animal (bite)	1	0 (0%)
Total		16	11 (69%)

Authors' own work.

muscle tendon and the tendon of the extensor digitorum muscle before its trifurcation—see Fig. 12.4 and the excellent animated gifs of the muscles (*Extensor digitorum longus muscle animated gif*, 2019; *Tibialis anterior muscle animated gif*, 2019). Thompson (Thompson, 1952) suggests this was an Achilles tendon injury. However, based on the translation below by Dr. Fabian Horn anterior leg has been depicted "Then fate snared Diores, the son of Amarynkeus: (4.517) For he was struck with a jagged stone by his ankle in his right lower leg. The leader of the Thracian men struck him, Peiros, the son of Imbrasos, who had come from Aenus. (520) Both sinews and the bones of the shameless stone did crush completely. And backward into the sands he fell, reaching out with both hands for his dear companions, breathing out his life. But the one who had struck him ran up, Peiros, and wounded him with his spear beside the navel, and out streamed (525) all his entrails to the ground, and darkness covered his eyes." The victim, Diores, was left unable to move when the rock injured and then killed by a spear into his belly (not shown).

Note 3

See Fig. 12.5. Ralph Thompson (1876–1955), a surgeon with a specialty in urology, noted that the spear or arrow had to pass through the sciatic notch, through the urinary bladder and then come out under the pubic arch. Thompson also commented that he doubted his surgical trainees could describe this anatomy when looking at a patient's entry wound in the buttock (Thompson, 1952). Saunders (1999), a physician classicist, does not disagree with Thompson's analysis and adds that the prepositions that Homer uses in his description are also consistent with Thompson's proposed path for the weapons (see Saunders, 1999 Figures 5 and 6).

This injury must have fascinated Homer since he described it twice in Table 12.1 (Cases 4 and 12). It certainly seems like something Homer must have seen in person,

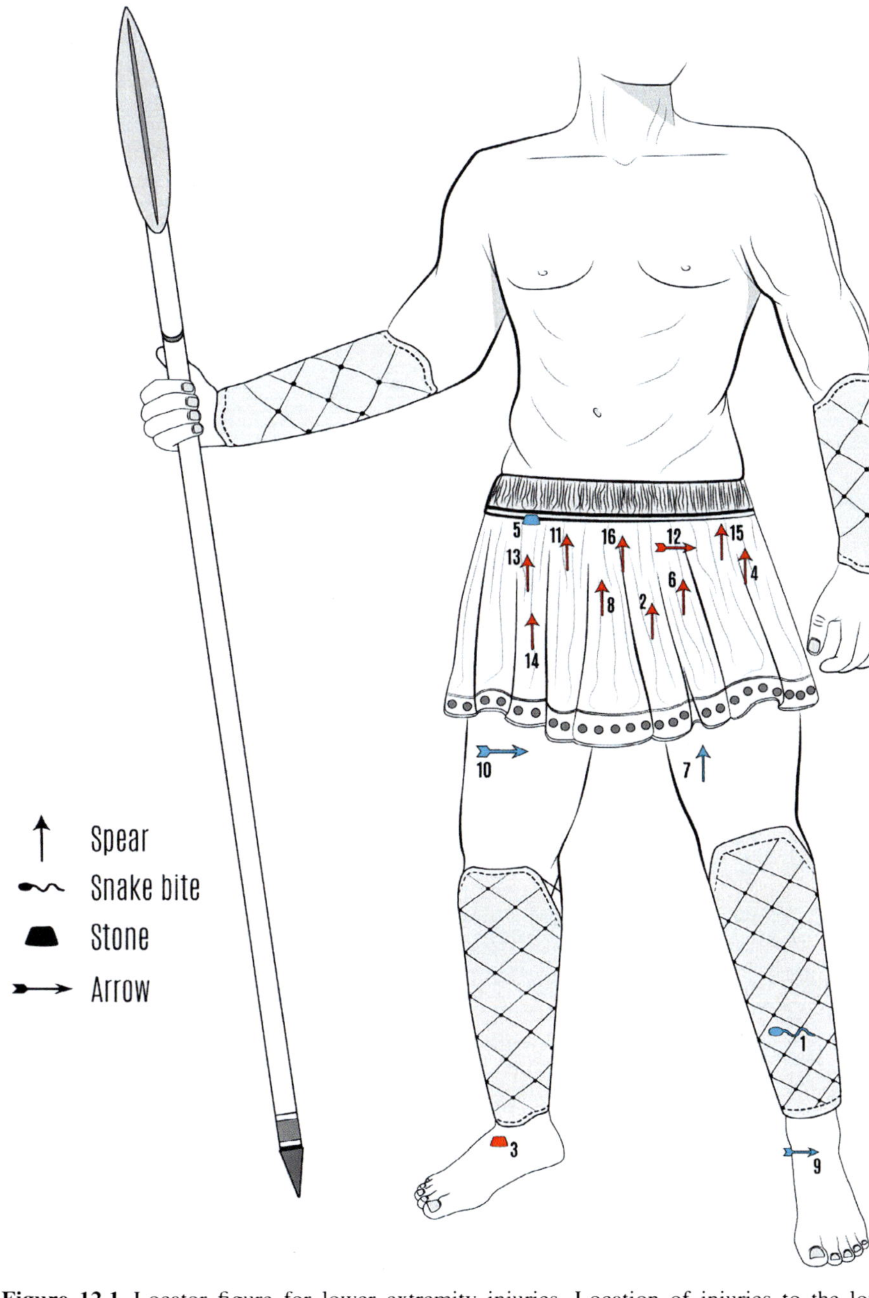

Figure 12.1 Locator figure for lower extremity injuries. Location of injuries to the lower extremity and pelvis in the *Iliad* as listed in Table 12.1 are shown diagrammatically. Symbols (see key in the figure) indicate the weapon used to cause the injury. Red indicates fatality, and blue indicates survival.

Source: Image by Mileidy Fernandez after an image drawn by Anusha Lekshminarayanan.

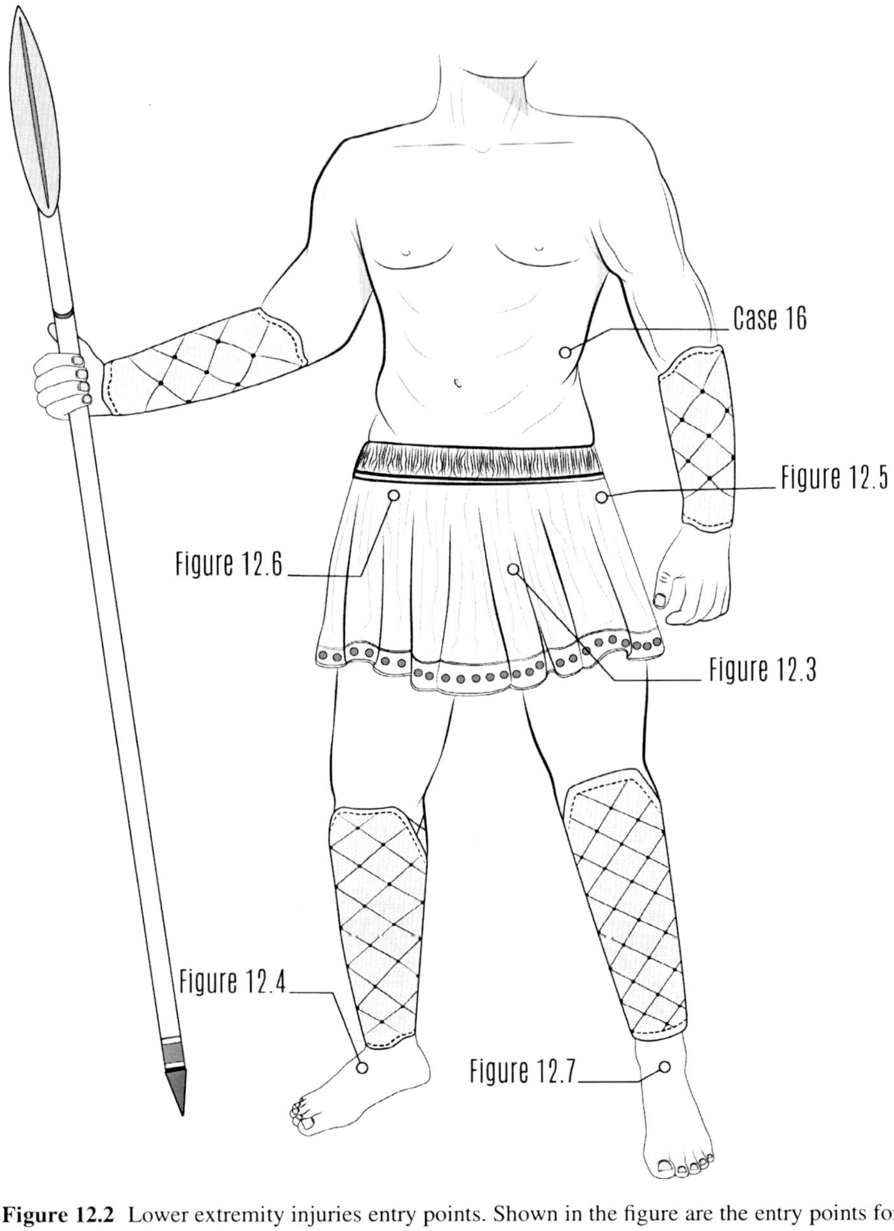

Figure 12.2 Lower extremity injuries entry points. Shown in the figure are the entry points for the injuries described in Cases 2 (6, 8, and 11), 3, 4, 7, and 16.
Source: Image by Mileidy Fernandez after an image drawn by Anusha Lekshminarayanan.

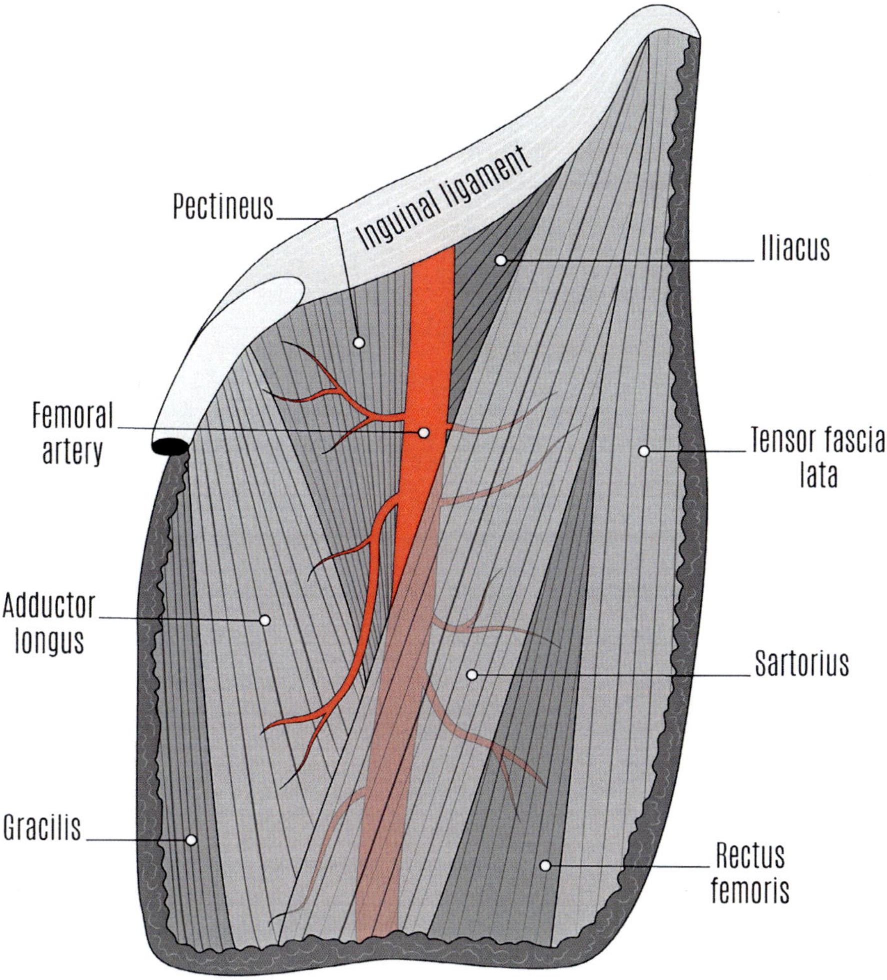

Figure 12.3 Femoral artery. A significant laceration of the femoral artery can cause fatal bleeding without prompt, effective treatment.
Source: Image by Mileidy Fernandez.

in particular, observing that patients were passing blood in the urine and concluding that the urinary bladder had been pierced.

Note 4

One of us (TL) has seen a number of individuals with gunshot wounds with acetabulum and femur injury without vascular injury who survived without surgery.

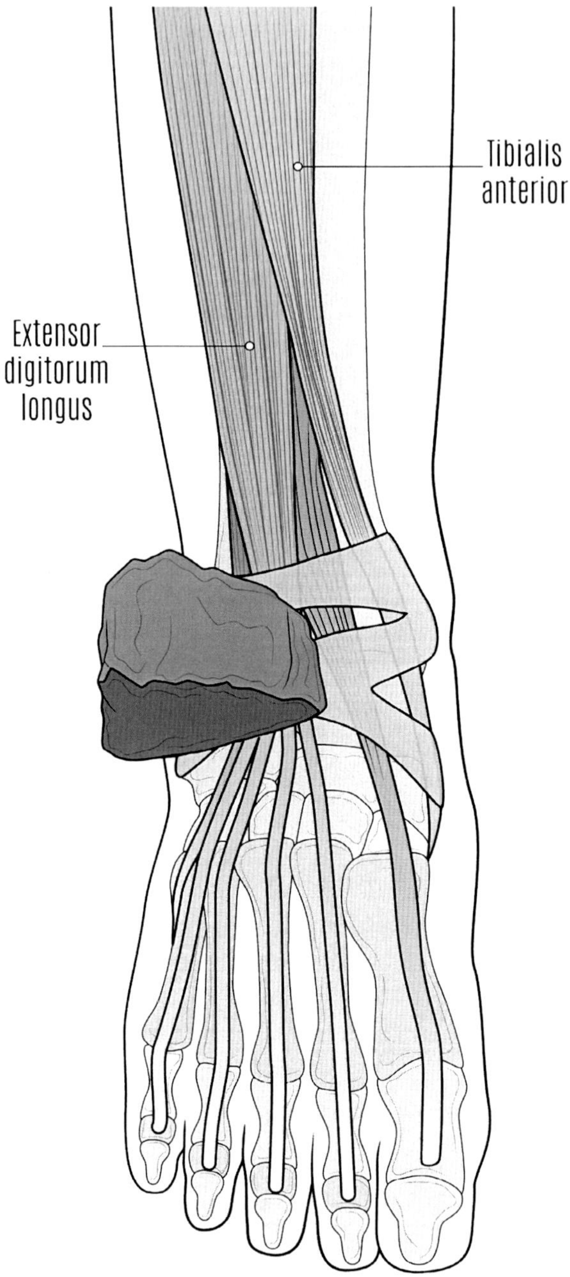

Figure 12.4 Anterior tibialis and extensor digitorum longus injuries revised. Suggested location for Diores' (Case 3 in Table 12.1) ankle injury. An injury at this location would cut the tendons of both the anterior tibialis muscle and extensor digitorum muscle (before its trifurcation). See text for further discussion.
Source: Image by Mileidy Fernandez after an image drawn by Anusha Lekshminarayanan.

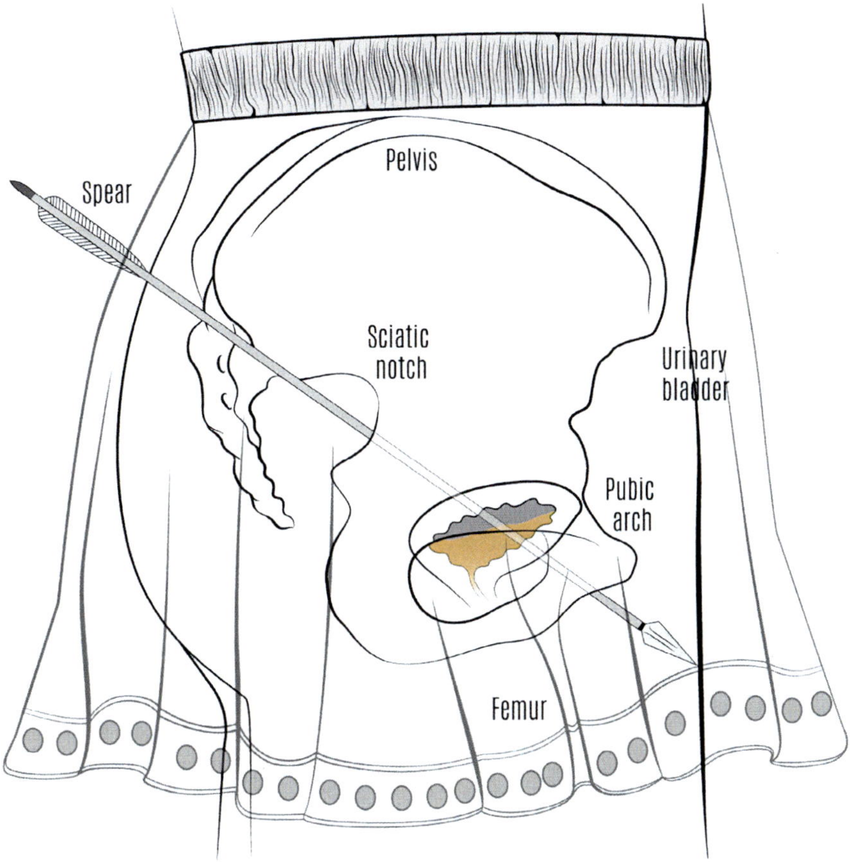

Figure 12.5 Spear buttock, bladder injury. Spear to the buttock exiting through the urinary bladder (Case 4, Table 12.1). See text for discussion.
Source: Image by Mileidy Fernandez after an image drawn by Anusha Lekshminarayanan.

So, even if, unlike Aeneas, your mother isn't a goddess[1] a proximal femur/acetabular fracture (see Fig. 12.6) with conscientious, careful, conservative care is survivable without surgical intervention. Indeed, interestingly, anthropologist Margaret Mead is

[1] Then, the warrior fell upon his knees, and thus abode, and with his stout hand leaned he upon the earth, and dark might enfolded his eyes And now would the king of men, Aeneas, have perished, had not the daughter of Zeus, Aphrodite, been quick to mark, even his mother, that conceived him to Anchises as he tended his knee. About her dear son, she flung her white arms, and before him she spread a fold of her bight garment to be a shelter against missiles, lest any of the Danaans with swift horses might haul a spear of bronze into his breast and take away his life. She then was bearing her dear son forth from out the battle. (*Iliad*, Book 5, lines 308–317, translation from Homer (1924). This work is licensed under a Creative Commons Attribution-ShareAlike 3.0 United States License.)

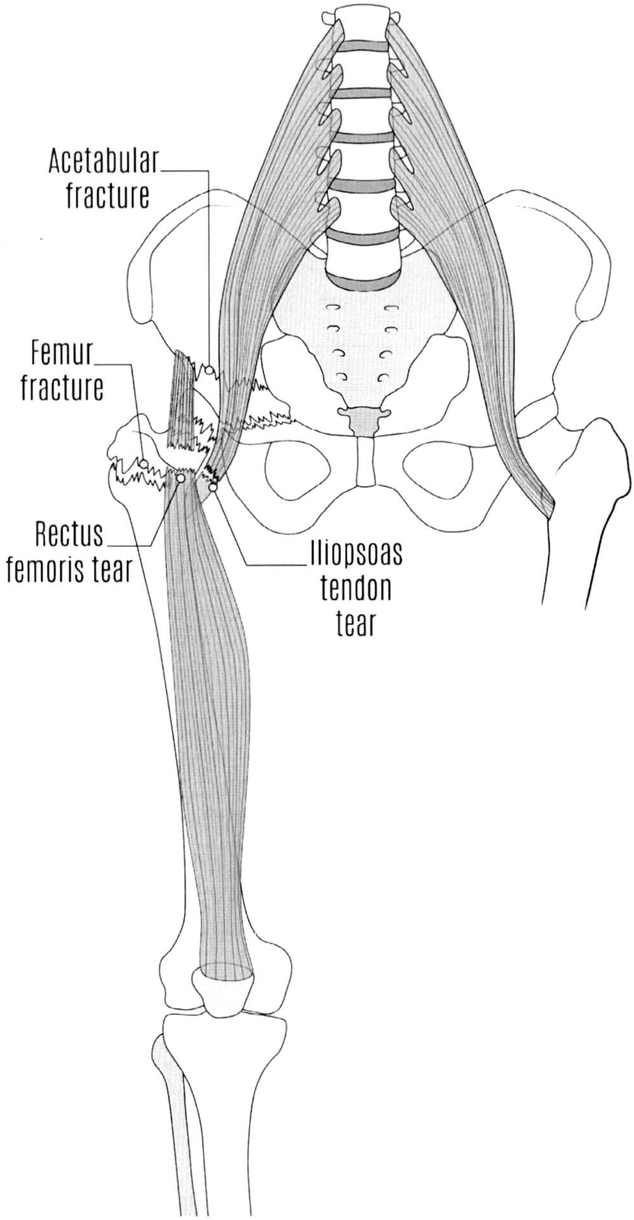

Figure 12.6 Aeneas' acetabular/femur fracture. Illustration of acetabular/femur fracture of Aeneas (Case 5 in Table 12.1). See text for further discussion.
Source: Image by Mileidy Fernandez after an image drawn by Anusha Lekshminarayanan.

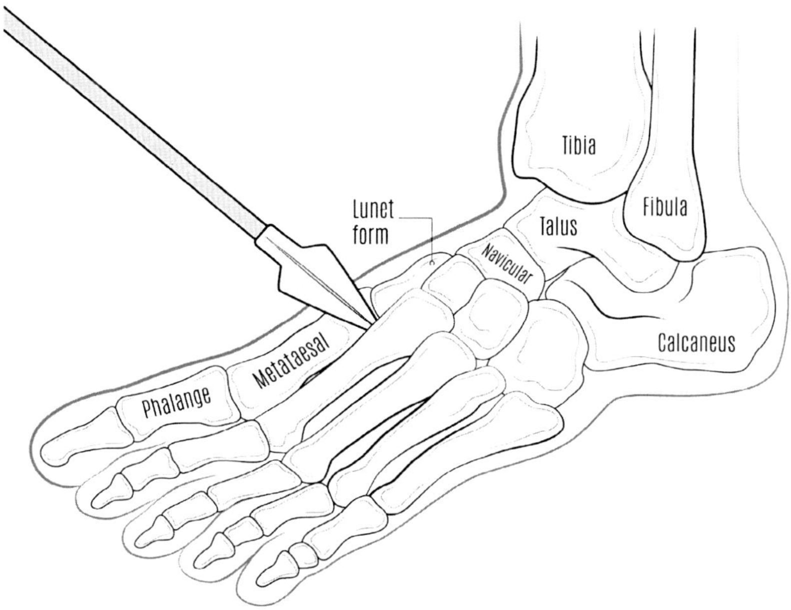

Figure 12.7 Arrow is shown landing on metatarsals. Alexandros' (Paris') arrow lands on the metatarsals of Diomedes' foot (Case 9 in Table 12.1).
Source: Image by Mileidy Fernandez after an image drawn by Anusha Lekshminarayanan.

reported to have said[2] that the first sign of civilization is not a harvested crop or first pot but a healed femur fracture: for, she reasoned, in the animal kingdom, if you break your leg, you die because you are unable to run from danger, hunt, and survive. But a healed thigh bone is a sign that someone has taken time to care for the injured (Margaret Mead "First Sign of Civilization," 2023).

Galenakos et al. (2015) comment that Aeneas injury is similar to the description in the article by Pearson and Hargadon (1962). Indeed, in Pearson's study of 80 cases of patients with acetabular floor fractures, almost all patients for whom follow-up was available, survived. Most survived and a large majority of patients without central dislocation of the femoral head had no or minimal signs or symptoms needing only conservative treatment—that is, rest for 1–6 weeks. Aeneas was functionally recovered and back fighting and inflicting death on Greek warriors the same day[3] (see Cases 40, Table 6.1) and 4 days later (Case 129, Table 6.1).

Diomedes' foot injury (Case 9, Table 12.1) is shown in Fig. 12.7.

We see that Homer shows excellent knowledge of the vascular, muscular, and skeletal structures of the lower limb (Galanakos et al., 2015; Kömürcü et al., 2014; Saunders, 1999; Thompson, 1952).

[2] Not without some disputation.

[3] This is consistent with the times course of events in the Iliad being contracted for dramatic effect, rather than the Epic being a purely historical recounting.

As noted in previous chapters, injuries to major arteries lead to rapid death in Homer's casualties. Injuries without clear vascular injuries result either in survival or delayed death. By what means did Homer acquire his excellent knowledge of anatomy? Most surely by treating multiple casualties and also likely by having access to bodies for post-mortem.

References

Battle of Shiloh. (2024). https://en.wikipedia.org/wiki/Battle_of_Shiloh.

Extensor digitorum longus muscle animated gif. (2019). https://commons.wikimedia.org/wiki/File:Extensor_digitorum_longus_muscle_-_animation_2.gif.

Galanakos, S. P., Bot, A. G. J., & Macheras, G. A. (2015). Pelvic and lower extremity injuries in Homer's Iliad: A review of the literature. *The Journal of Trauma and Acute Care Surgery, 78*(1), 204–208. https://doi.org/10.1097/TA.0000000000000486.

Gitto, L., Bonaccorso, L., & Serinelli, S. (2019). Death due to severe blood loss following an accidental lesion to the femoral vessels. *The Medico-Legal Journal, 87*(4), 196–201. https://doi.org/10.1177/0025817219875425.

Heard, L. (1862). Wound of the femoral artery—Ligature of both cardiac and distal sides of injury—Secondary hæmorrhage—Ligature of external iliac—Peritonitis—Death of the patient. *The Boston Medical and Surgical Journal, 67*(19), 369–373. https://doi.org/10.1056/NEJM186212110671901, https://www.nejm.org/doi/full/10.1056/NEJM186212110671901.

Homer. (1924). *The Iliad*. Harvard University Press.

Johnson, H. A. (2003). The foot that stalled a thousand ships: A controversial case from the 13th century BCE. *Journal of the Royal Society of Medicine, 96*(10), 507–508.

Kömürcü, E., Tok, F., Simşek, A., & Ozçakar, L. (2014). Musculoskeletal injuries in Homer's Iliad: The War of Troy revisited. *American Journal of Physical Medicine & Rehabilitation, 93*(4), 335–341. https://doi.org/10.1097/PHM.0000000000000051.

Margaret Mead 'First Sign of Civilization'. (2023). https://www.snopes.com/fact-check/margaret-mead-healed-femur/.

Pearl, J. P., McNally, M. P., & Perdue, P. W. (2003). Femoral vessel injuries in modern warfare since Vietnam. *Military Medicine, 168*(9), 733–735.

Pearson, J. R., & Hargadon, E. J. (1962). Fractures of the pelvis involving the floor of the acetabulum. *The Journal of Bone and Joint Surgery. British Volume, 44-B*, 550–561.

Saunders, K. B. (1999). The wounds in "Iliad" 13-16. *Classical Quarterly, 49*(2), 345–363.

Thompson, A. R. (1952). Homer as a surgical anatomist. *Proceedings of the Royal Society of Medicine, 45*(11), 765–767.

Tibialis anterior muscle animated gif. (2019). https://commons.wikimedia.org/wiki/Category:Tibialis_anterior_muscles#/media/File:Tibialis_anterior_muscle_-_animation.gif.

Cranio-maxillofacial injuries in Homer

13

Tyler Lucas[1], Anusha Lekshminarayanan[2,3], and Eric L. Altschuler[4,5]
[1]Division of Orthopaedic Surgery, Department of Surgery, Metropolitan Hospital, New York, NY, United States, [2]Department of Physical Medicine and Rehabilitation, Westchester Medical Center, Valhalla, NY, United States, [3]Department of Rehabilitation Medicine, New York Medical College, Valhalla, NY, United States, [4]Essen Health Care, Bronx, NY, United States, [5]Kessler Institute for Rehabilitation, West Orange, NJ, United States

Cranio-maxillofacial injuries in Homer's *Iliad* are cataloged in Tables 13.1 and 13.2. Homer was clearly well aware that strikes to the head and neck were an extremely lethal and efficient way to kill an adversary (Kömürcü et al., 2014; Mylonas et al., 2008; Sapounakis et al., 2007). Fatality rates and weapons used are listed in Table 13.3. The site of injuries of Greek warriors is shown schematically in Fig. 13.1 and those suffered by Trojan warriors in Fig. 13.2.

For most of the injuries in Tables 13.1 and 13.2, the mechanism of instant death is obvious from the location and weapon of injury without any specialized medical knowledge. However, a couple of the injuries merit further discussion.

Note 1

Case 4 from Table 13.1 (Book 5, lines 290–295):

(Translation from Homer (1924), this work is licensed under a Creative Commons Attribution-ShareAlike 3.0 United States License.)

[290] So spake he [Diomedes] and hurled; and Athene guided the spear upon his [Pandarus'] nose beside the eye, and it pierced through his white teeth. So the stubborn bronze shore off his tongue at its root, and the spear-point came out by the base of the chin. Then he fell from out the car, [295] and his armor all bright and flashing clanged upon him.

To have been able to describe the injury in such detail—the spear entering at the nose, then going through the teeth, then the tongue before finally coming out at the chin—Homer must have observed the injury up close. Indeed, this description is a time machine taking you back to view the gruesome injury with Homer.

Note 2

Here is the description of Thoön's injury (Book 13, lines 545–549):

[545] Then Antilochus, biding his time, leapt upon Thoön, as he turned his back, and smote him with a thrust, and wholly severed the vein that runneth along the back

Medicine in Homer. DOI: https://doi.org/10.1016/B978-0-443-32978-4.00010-9

Table 13.1 Cranio-maxillofacial injuries in the *Iliad* Books 1–12 (*G*, Greek; *T*, Trojan).

Case number	Book;line	Location of injury	Victim	Attacker	Weapon	Outcome	Notes
1	4;457–462	Forehead	Echepolus (T)	Antilochus (G)	Spear	Death	
2	4;494–504	Temple	Democoon (T)	Odysseus (G)	Spear	Death	
3	5;72–75	Mouth	Pedaeus (T)	Meges (G)	Spear	Death	
4	5;290–293	Nose	Pandarus (T)	Diomedes (G)	Spear	Death	See Note 1.
5	5;584–586	Temple	Mydon (T)	Antilochus (G)	Sword	Death	
6	5;655–658	Neck	Tlepolemus (G)	Sarpedon (T)	Spear	Death	
7	6;5–11	Forehead	Acamas (T)	Aias (G)	Spear	Death	
8	7;11–12	Neck	Eioneus (G)	Hector (T)	Spear	Death	
9	7;260–262	Neck (grazed)	Hector (T)	Aias (G)	Spear	Survival	
10	10;454–457	Neck	Dolon (T)	Diomedes (G)	Sword	Death	
11	11;91–98	Forehead	Oileus (T)	Agamemnon (G)	Spear	Death	
12	11;107–109	Ear	Antiphus (T)	Agamemnon (G)	Sword	Death	
13	11;145–147	Decapitation	Hippolochus (T)	Agamemnon (G)	Sword	Death	
14	11;238–240	Neck	Iphidamas (T)	Agamemnon (G)	Sword	Death	
15	11;256–261	Decapitation	Coon (T)	Agamemnon (G)	Spear	Death	
16	12;182–186	Forehead	Damascus (T)	Polypoetes (G)	Spear	Death	
17	12;378–386	Head	Epicles (T)	Aias (G)	Rock	Death	

Source: Adapted from Tables 4 and 5 in Mylonas et al. (2008), used with permission.

Table 13.2 Cranio-maxillofacial injuries in the *Iliad* Books 13–23 (*G*, Greek; *T*, Trojan).

Case number	Citation	Injuries	Victim	Attacker	Weapon	Outcome	Notes
18	13;177–18113: 202–205	Beneath the ear Decapitation	Imbrius (T) Imbrius (T)	Teucer (G)Aias (G)	Spear Sword	Death Decapitation of corpse	
19	13; 387–391	Anterior neck (Zone II)	Asius (T)	Idomeneus (G)	Spear	Death	
20	13; 541–544	Anterior neck (Zone II)	Aphareus (G)	Aeneas (T)	Spear	Death	
21	13: 545–549	Posterior neck	Thoön (T)	Antilochus (G)	"thrust" [weapon no specified, spear presumed]	Death	See Note 2.
22	13;576–580	Temporal bone	Deipyrus (G)	Helenus (T)	Sword	Death	
23	13;614–618	Glabella	Peisander (T)	Menelaus (G)	Sword	Death	
24	13;671–672	Lateral neck (Zone III)	Euchenor (G)	Paris (T)	Arrow	Death	
25	14;465–468	Neck (Zone III)	Archelochus (T)	Aias (G)	Spear	Death	
26	14;492–500	A. Spearing followed by B. fatal Decapitation	Ilioneus (T)	Peneleos (G)	A. spear B. Sword (shaft)	A. Severe brain injury B. Death	See Chapter 15 for further discussion of this injury.
27	15;429–435	Temporal and some parts of parietal areas of the cranium	Lycophron (G)	Hector (T)	Spear	Death	

(*Continued*)

Table 13.2 (Continued)

Case number	Citation	Injuries	Victim	Attacker	Weapon	Outcome	Notes
28	15;449–453	Posterior neck	Cleitus (T)	Teucer (G)	Arrow	Death	
29	16;330–334	Neck	Cleobulus (T)	Aias (G)	Sword	Death	
30	16;339–341	Lateral neck (Zone III)	Lyco (T)	Peneleos (G)	Sword	Death	
31	16;345–350	Oral cavity and basion	Erymas (T)	Idomeneus (G)	Spear	Death	
32	16;401–410	Right mandible ramus and teeth	Thestor (T)	Patroclus (G)	Spear	Death	
33	16;411–414	Head (split in two)	Erylaus (T)	Patroclus (G)	Stone	Death	
34	16;577–580	Head (split in two)	Epeigeus (T)	Hector (G)	Stone	Death	
35	16;586–587	Neck	Sthenelaus (T)	Patroclus (G)	Stone	Unknown	
36	16;603–607	Lateral neck (Zone III)	Laogonus (T)	Meriones (G)	Unknown	Death	
37	16;733–743	Forehead	Cebriones (T)	Patroclus (G)	Stone	Death	
38	17;45–50	Anterior neck (Zone II)	Euphorbus (T)	Menelaus (G)	Spear	Death	
39	17;293–303	Head	Hippothous (T)	Aias (G)	Spear	Death	
40	17;615–619	Lateral neck (zone II) through teeth and tongue	Coeranus (T)	Hector (G)	Spear	Death	
41	20;386–388	Head (split in two)	Iphition (T)	Achilles (G)	Spear	Death	

Table 13.2 (Continued)

Case number	Citation	Injuries	Victim	Attacker	Weapon	Outcome	Notes
42	20;395–400	Temporal bone	Demoleon (T)	Achilles (G)	Spear	Death	
43	20;455–456	Neck	Dryops (T)	Achilles (G)	Spear	Death	
44	20;472–474	Ear/Pinna	Mulius (T)	Achilles (G)	Spear	Not stated but presumed death.	Then with his spear Achilles drew nigh unto Mulius and smote him upon the ear, and clean through the other ear passed the spear-point of bronze
45	20;474–477	Head	Echeclus (T)	Achilles (G)	Sword	Death	
46	20;481–483	Decapitation	Deucalion (T)	Achilles (G)	Sword	Death	
47	22;324–327	Anterior neck (Zone II)	Hector (T)	Achilles (G)	Spear	Death	
48	23;689–691	Cheek	Euryalus (G)	Epeius (G)	Fist	Survival	

Source: Adapted from Tables 5–8 in Mylonas et al. (2008), used with permission.

Table 13.3 Weapons used to inflict cranio-maxillofacial injuries in the *Iliad*.

Weapon number	Weapon used	Number of injuries	Number fatal (%)
1	Spear	28	26 (93%)
2	Sword	12	12 (100%)
3	Stones	4	3 (75%)
4	Arrow	2	2 (100%)
5	Rocks	1	1 (100%)
6	Unknown	1	1 (100%)
7	Fist	1	0 (0%)
Total		49	45 (92%)

Source: Adapted from Table 13.3 in Mylonas et al. (2008), used with permission.

continually until it reacheth the neck; this he severed wholly, and Thoön fell on his back in the dust, stretching out both his hands to his dear comrades.

Thompson a general surgeon and urologist, notes that this is the only time Homer mentions a vein and suggests (Thompson, 1952) that the vein could refer to the chain of the right innominate vein, the superior vena cava, the right auricle, and the inferior vena cava (though some of these vessels are anterior). Saunders, a physician and classicist, says there is no such vein consistent with Homer's description (Saunders, 1999) and specifically rules out the aorta (an artery) the vena cava and the internal jugular vein. Neither of these authors nor any others to our knowledge have considered that Antilochus lacerated the *posterior branch* of Thoön's *external* jugular vein (Fig. 13.3, Posterior external jugular vein, 2024). This vein fits anatomically with the text, and also, if severed, could produce a rapidly fatal injury (Jones, 2023) with dramatic effect as described by Homer.

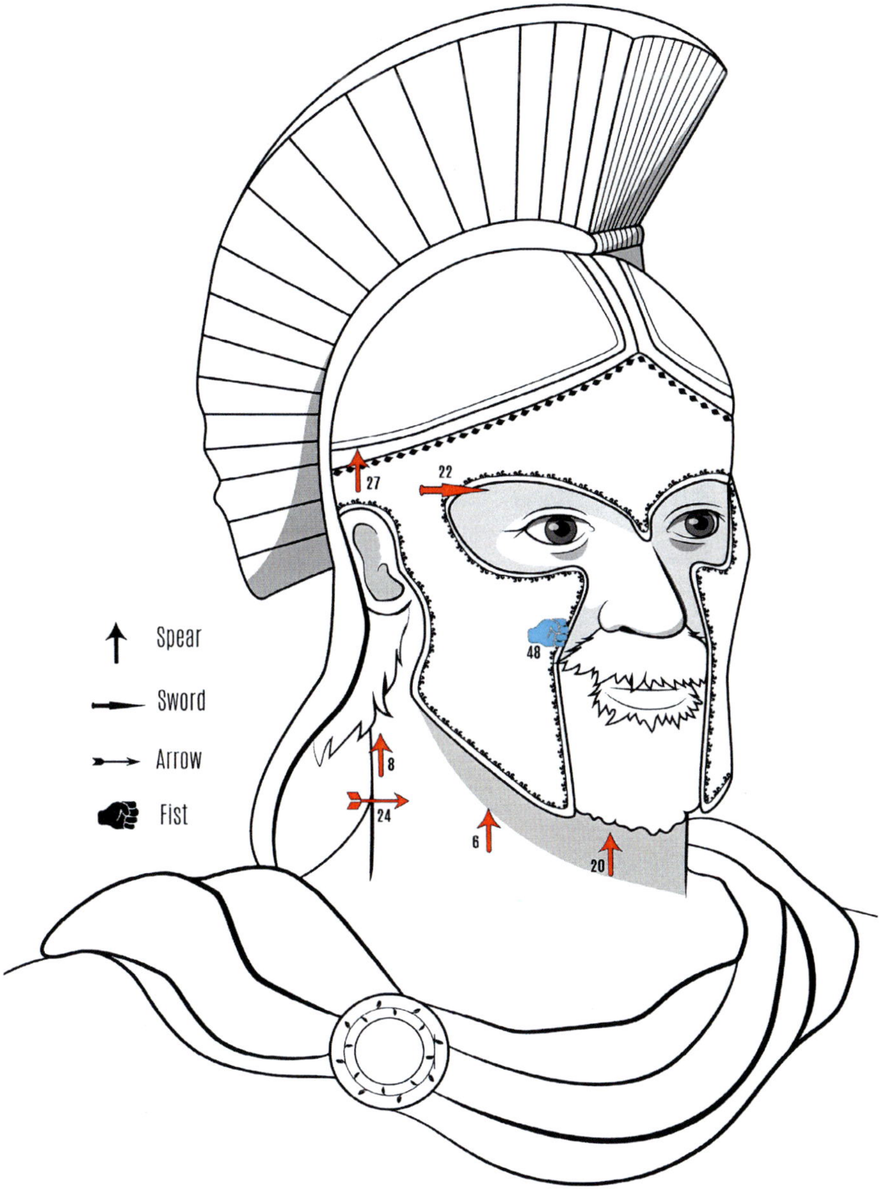

Figure 13.1 Greek cranio-facial injuries. Cranio-maxillofacial injuries of Trojan warriors listed in Tables 13.1 and 13.2. Red denotes fatality and blue survival.
Source: Image by Mileidy Fernandez after an image drawn by Anusha Lekshminarayanan.

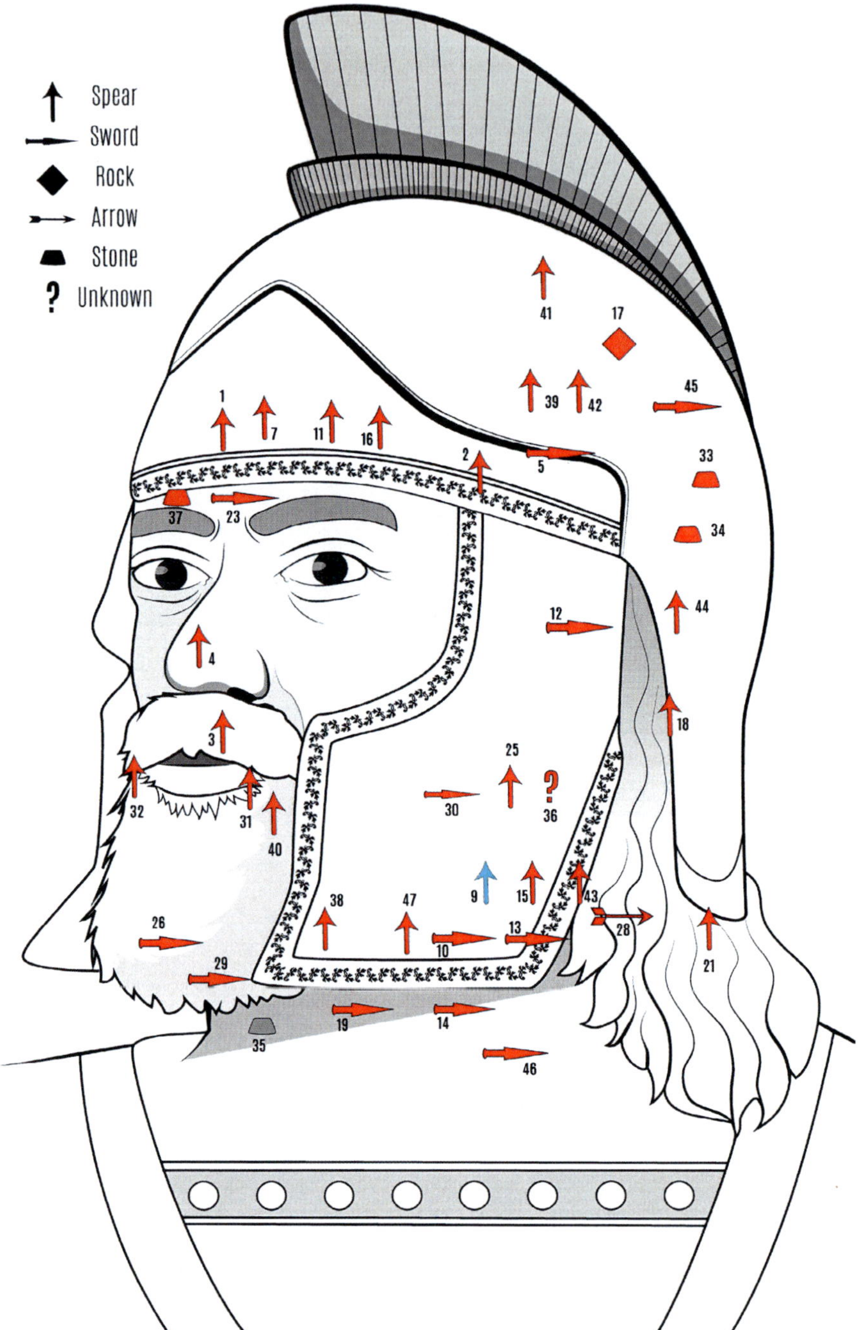

Figure 13.2 Trojan cranial-facial injuries. Cranio-maxillofacial injuries of Trojan warriors listed in Tables 13.1 and 13.2. Red denotes fatality, blue survival, and gray unknown outcome. *Source*: Image by Mileidy Fernandez after an image drawn by Anusha Lekshminarayanan.

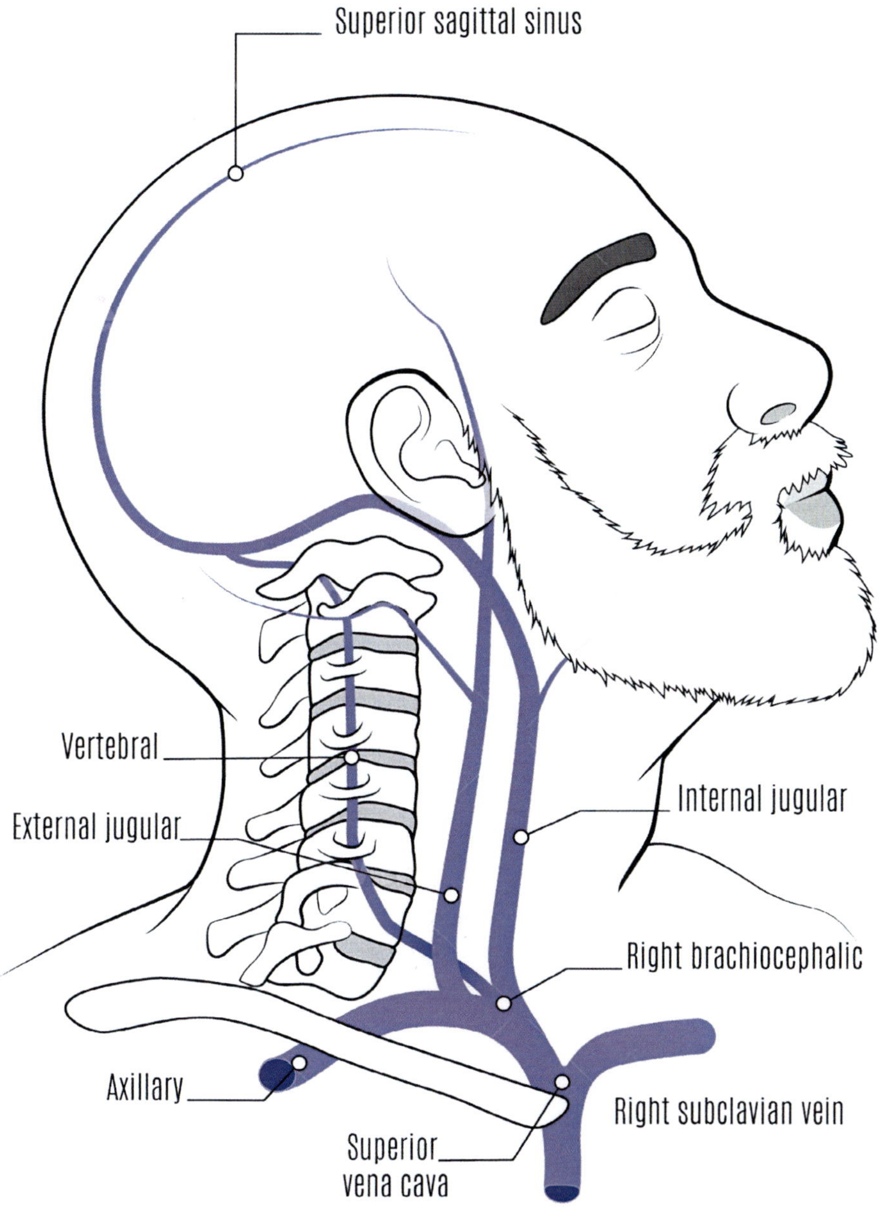

Figure 13.3 External jugular vein. Veins of the posterior head and neck.
Source: Image by Mileidy Fernandez after an image By Henry Vandyke Carter–Henry Gray (1918) Anatomy of the Human Body (See "Book" section below) Bartleby.com: Gray's Anatomy, Plate 563, Public Domain, https://commons.wikimedia.org/w/index.php?curid=522459.

References

Homer. (1924). *The Iliad*. Harvard University Press.

Jones, O. (2023). Posterior triangle of the neck. https://teachmeanatomy.info/neck/areas/posterior-triangle/.

Kömürcü, E., Tok, F., Simşek, A., & Ozçakar, L. (2014). Musculoskeletal injuries in Homer's Iliad: The War of Troy revisited. *American Journal of Physical Medicine & Rehabilitation, 93*(4), 335–341. https://doi.org/10.1097/PHM.0000000000000051.

Mylonas, A. I., Tzerbos, F. H., Eftychiadis, A. C., & Papadopoulou, E. C. (2008). Cranio-maxillofacial injuries in Homer's Iliad. *Journal of Cranio-Maxillo-Facial Surgery, 36*(1), 1–7. https://doi.org/10.1016/j.jcms.2007.07.010.

Posterior external jugular vein. (2024). https://en.wikipedia.org/wiki/Posterior_external_jugular_vein.

Sapounakis, C., Rallis, G., Mourouzis, C., Konsolaki, E., & Tesseromatis, C. (2007). Injuries to the head and neck in Homer's Iliad. *The British Journal of Oral & Maxillofacial Surgery, 45*(2), 112–115.

Saunders, K. B. (1999). The wounds in "Iliad" 13–16. *Classical Quarterly, 49*(2), 345–363.

Thompson, A. R. (1952). Homer as a surgical anatomist. *Proceedings of the Royal Society of Medicine, 45*(11), 765–767.

Trauma in Homer: a classicist's analysis and perspective

14

Fabian Horn
Ludwig-Maximilians-University Munich, Department of Greek and Latin Philology, Munich, Germany

The *Iliad* and *Odyssey*, which have come down to us under the name "Homer," even though we know nothing about their poet(s) as a historical person,[1] are Ancient Greek epic poems from the narrative sequence of the so-called Epic Cycle, the traditional Greek stories around the Trojan War. The historicity of the Trojan War is a matter of ongoing debate, but the historical event it might have been based on would have happened in the late Mycenaean Bronze Age about mid-12th century BCE, while the Homeric poems in the shape as they have been transmitted to us are a product of the mid- to late 8th century BCE. The Homeric epics are not works of history, but poetry originally sung and performed by traveling bards. The language and diction of the epics developed over the span of centuries and were shaped by the necessities of their meter, the dactylic hexameter. As parts of the epic narrative of the Greek past, the *Iliad* is set in the last year of the Greeks' ten-year-long siege of Troy, while the *Odyssey* is the story about Odysseus' return to his home in Ithaca after the conclusion of the war, finally culminating in the reassertion of his rule over his domain. As such, the *Iliad* is a poem full of war and battles, and to a much lesser degree, so is the *Odyssey*.[2]

As a poem of war, long passages of the *Iliad*, particularly the so-called "battle books" (*Iliad* books 5–8, 11–17, and 20–22), consist of little other than descriptions of fighting, while Odysseus' battles in the *Odyssey* are confined to book 22, where he dispatches the 108 suitors of his wife Penelope. Both poems belong to the same tradition of oral epic poetry and therefore fighting, wounding, and dying are expressed

[1] Date and authorship of the Homeric poems are notoriously fraught with controversy to the extent that the uncertainties about the poems' composition are commonly referred to as "the Homeric question" among Classical scholars. While the question of the date of composition and whether *Iliad* and *Odyssey* were works of the same poet are not of importance for the present argument, I do assume that the *Iliad* was ultimately "masterminded" by one monumental poet who either composed the whole poem in the traditional oral style of epic composition or was responsible for some form of final redaction on the basis of existing material. Hence, I am using the traditional name "Homer" as a convenient shorthand for "the poet of the *Iliad*."

[2] The poems have been divided into 24 "books" by Alexandrinian scholars (the precise number of letters of the Greek alphabet), according to which both the *Iliad* and the *Odyssey* are conventionally cited, the *Iliad* with upper-case Greek letters and the *Odyssey* with lower-case letters. However, for the sake of convenience I will use Arabic numerals in the following chapter, prefaced by *Il.* for the *Iliad* or *Od.* for the *Odyssey*.

Medicine in Homer. DOI: https://doi.org/10.1016/B978-0-443-32978-4.00020-1

in a formulaic, stylized manner. Homeric battles only rarely present strategic maneuvers of contingents of warriors, but almost exclusively consist of concatenations of individual battle encounters between two fighters. The techniques and historical realities of Homeric warfare have been examined and discussed by historians (see, e.g., Albracht, 2005; van Wees, 1986, 1988, 1994), but for philologists the important feature of how battles are depicted in Homeric poetry is that fights between individual heroes are foregrounded (see, e.g., Marg, 1976), which have aptly been characterized as "Homeric duels" (e.g., Gazis, 2023). These duels are usually rather brief with hardly ever a longer exchange of blows. Their description is stylized and the minimal form of such a duel is the naming of the slayer and the slain with a brief reference either to the act of killing or the act of dying, without any further details. Even in cases where death is not explicitly indicated, the wounded warriors usually do not reappear and the description of the mortal blow or the fall of the struck warrior might metonymically signify the occurrence of death (see Horn, 2018, 362–368; Saunders, 2003, 132–134 notes that even minor injuries often cause death but is undecided in cases where death is not explicitly indicated). In the *Odyssey*, there are no nonlethal injuries at all and in the *Iliad* it is a comparatively rare occurrence that one of the warriors engaged in a fight on the battlefield is nonfatally injured and withdraws to recover and to return to the fight later. As a result, descriptions of the treatment of wounds and medicine in the *Iliad* are comparatively few, but these have also been studied by philologists and historians of medicine (for discussions see, e.g., Frölich, 1879; Körner, 1929, 70–86; Majno, 1975, 142–147; Salazar, 2000, 126–158; Lane Fox, 2020, 11–22).

The stylization and poetic presentation of Homeric battle scenes become especially apparent in what is missing from them: In a Bronze Age conflict, such as the Trojan War would have been, it seems unlikely that most wounds struck would prove immediately fatal (as it is depicted in the *Iliad*), and historically fighters would have also suffered wounds, amputations, or disfiguring injuries. However, the lack of nonfatal injuries in Iliadic battle scenes is not a result of the Homeric warriors' particular deadliness but needs to be attributed to the narrative function of being only wounded rather than killed. Given the danger of fighting on the battlefield and the high mortality rate of wounds (see detailed Saunders, 2004, 6–13), not being killed but nevertheless receiving a wound is the sign of important heroes which illustrates their power, resilience, or divine favor (see esp. Neal, 2006). Therefore being wounded is an honor which is only accorded to major heroes, such as Menelaos (*Il.* 4.124–219), Diomedes (*Il.* 5.98–143; 11.375–400), Hector (*Il.* 7.260–272; 14.409–439), Sarpedon (*Il.* 11.660–698), Agamemnon (*Il.* 11.252–283), Odysseus (*Il.* 11.434–488), Machaon (*Il.* 11.505–520), or Achilles (*Il.* 21.166–167). Achilles' wound best illustrates the principle, since he does not even seem to register that he was wounded and without flinching proceeds to kill Asteropaeus (text: van Thiel, 2010; translation: Horn):

Il. *21.162–170: (…) ὁ δ' ἁμαρτῇ δούρασιν ἀμφις*
ἥρως Ἀστεροπαιος, ἐπει περιδέξιος ἦεν·
καί ῥ' ἑτέρῳ μὲν δουρι σάκος βάλεν, οὐδὲ διὰ πρὸ
ῥῆξε σάκος· χρυσὸς γὰρ ἐρύκακε, δῶρα θεοιο·

τῷ δ' ἑτέρῳ μιν πῆχυν ἐπιγράβδην βάλε χειρὸς
δεξιτερῆς, σύτο δ' αιμα κελαινεφές· ἣ δ' ὑπὲρ αὐτοῦ
γαίῃ ἐνεστήρικτο λιλαιομένη χροὸς ἆσαι.
δεύτερος αὖτ' Ἀχιλεὺς μελίην ἰθυπτίωνα
Ἀστεροπαίῳ ἐφῆκε κατακτάμεναι μενεαίνων.

(...) but he took up both spears at the same time,
the hero Asteropaeus, for he was ambidextrous.
And with one spear he struck the shield, but it did not
break through the shield: for the gold warded it off, the gift of the god;
and with the other he struck a glancing blow to the ulna
of the right arm, and dark blood rushed out: but the spear point
passed over him and stuck in the earth, yearning to gut itself with his flesh.
In turn, Achilles threw second his straight-flying ashen spear
at Asteropaeus, eager to kill him.

Nonfatal injuries in the *Iliad* are rare and either insignificant, as in the example above, or quickly healed through divine intervention, so they never leave permanent damage. Similarly, mutilation in Homer only occurs *post mortem*, when warriors mistreat the corpse of a fallen enemy in anger, revenge, or as a means of deterrence (e.g., *Il.* 11.145–147; 13.201–205; 14,496–500; 22.395–405). It is presumably also due to poetic stylization that there are never any mutilated warriors depicted, since disfigurement would conflict with the image of the traditionally handsome and noble hero.

Thus, with the exception of the few minor wounds, most Iliadic encounters are decisive and leave one of the combatants dead—hence their designation as ἀνδροκτασίη, lit. "man-slaughter." In total, there are more than 240 such individual battle encounters resulting in immediate death of one of the combatants in the *Iliad* (see esp. Armstrong, 1969; Stoevesandt, 2004, 389–412), while in the *Odyssey* there is only one extended battle scene in *Od.* 22. Though there certainly are serious traumata which would cause instant death (such as decapitation), even when a wound is not so grave that it should result in "sudden death," warriors are usually said to die immediately after being struck. Kenneth Saunders, a Homeric scholar as well as a physician and physiologist, noted that "[v]ery few wounds in the *Iliad* would cause sudden death, yet most Homeric fighters seem to die that way or very rapidly. It is simply a convention (...)" (Saunders, 2003, 132). Thus, classical scholars have concluded that "[t]he Homeric poems are interested in death far more than they are in fighting" (Griffin, 1980, 94). This prevalence of (rapid rather than protracted) death as the result of almost all Iliadic duels is a stylization that can be attributed to the importance of dispatching enemies as a means of winning status in Homeric society. Hence, the fighters in the *Iliad* can be distinguished into two categories, the "major heroes," who perform feats of glory by killing series of opponents (while themselves suffering only nonlethal injuries), and the "small fighter," whose only function within the narrative is to be rapidly killed by a superior warrior as soon as they appear (see esp. Strasburger, 1954).

Despite the obvious importance of death, for the entertainment of the audience fights are frequently extended with additional elements such as flyting exchanges prior to the fight, details of the battle (weapon, mode of attack, location of the wound), an obituary or a simile for the dying warrior, or the subsequent despoliation of the corpse. As a result of the oral composition and performance of Archaic Greek epic poetry, Homeric battle descriptions follow a set pattern that still allows for sufficient flexibility and variation to be entertaining for the audience. The poetic conventions and the formulaic technique of these epic scenes of combat have been studied in great detail by philologists (e.g., Fenik, 1968; Friedrich, 2003; Hellmann, 2000; Horn, 2018; Morrison, 1999; Niens, 1987; Strasburger, 1954), who traditionally have been less interested in the specifics of the wounds inflicted in Homeric duels. Thus the granular investigations of injuries in Iliadic battles have been carried out mostly by physicians with an interest in Homeric poetry and medical historians (see, e.g., Nomikos, 2018; Saunders, 1999; of specific injuries see, e.g., Kömürcu et al., 2014: musculoskeletal injuries; Apostolakis et al., 2010; Santos, 2000: chest; Hutchison & Hirthler, 2013; Lipshutz, 1963: upper extremities; Kayhanian & Machado, 2020; Mylonas et al., 2008; Ralli et al., 2015; Sapounakis et al., 2007; head, face, and neck; on the *Odyssey* see Koutserimpas & Samonis, 2016).

Since the *Iliad* is neither a precise eye-witness account of historical events nor a medical textbook but rather a work of oral poetry, its representations of battles are subject to stylization and certain poetic conventions which seem to follow clear objectives. Therefore the questions traditionally posed by most philologists when reading epic battles scenes focus not so much on the medical details of the wounds described in individual battle encounters (since Classicists also often lack the necessary anatomical knowledge) but rather on the general meaning of the scenes within the narrative of the poem. The description of battles has been shown to be a means of exalting superior warriors by having them exhibit their strength and excellence in dispatching many opponents while emerging unscathed, or only with minor injuries (see, e.g., Hellmann, 2000, esp. 171–189; Horn, 2014, 64–98). The fact that in total more Trojans are killed than Greeks also foreshadows the final triumph of the Greeks and the fall of Troy (see Stoevesandt, 2004).

Yet philologists have also noted that despite all stylization there is a wide variety regarding the details of individual battle scenes. Even though fights are generally short and mostly finished after the first hit, the wounds that bring about the death of the inferior warriors are often described in minute detail. The anatomical elaborations sometimes added by the poet render the battle scenes not only vivid but also gruesome and realistic (cf. Friedrich, 2003: "biotic realism"/"truth to life"). Indeed, the amount of details provided in many cases has even led to the assertion that the poet Homer must have been a military physician (Frölich, 1879, 63–65; on Homer's anatomical knowledge also see Robertson, 2002; Thompson, 1952). However, closer study from scholars with medical expertise has also revealed that not all descriptions can be considered entirely factual and that the realism of the duels varies. This is of interest to Classicists insofar as these findings implicitly corroborate the current view of the *Iliad*'s oral composition and the intention of the poet as he sang the battle scenes. To illustrate these observations, I will proceed to present some instances of unrealistic battle descriptions.

The following example is a rather typical battles scene in providing details about weapon, wound, and the fall of the dying warrior, but it is unrealistic in that the weapon used could hardly cause the wound as described (text: van Thiel, 2010; translation: Horn):

Il. *16.321–325 (...) τοῦ δ' ἀντίθεος Θρασυμήδης*
ἔφθη ὀρεξάμενος πριν οὐτάσαι, οὐδ' ἀφάμαρτεν,
ὦμον ἄφαρ· πρυμνὸν δὲ βραχίονα δουρὸς ἀκωκὴ
δρύψ' ἀπὸ μυώνων, ἀπὸ δ' ὀστέον ἄχρις ἄραξε·
δούπησεν δὲ πεσών, κατὰ δὲ σκότος ὄσσε κάλυψεν

(...) godlike Thrasymedes
got the drop on him before he (i.e. Maris) could wound him, and he did not miss,
striking his shoulder quickly: and the point of the spear
sheared away from the muscles the uppermost part of the arm and tore off the bone entirely:
and he made a loud noise as he fell, and darkness fell over his eyes.

The fatal injury here is the loss of the arm, which has been cut off at the shoulder. The half-line ἀπὸ δ' ὀστέον ἄχρις ἄραξε (*Il.* 16.324) could mean both "he tore off the bone entirely" or "he completely shattered the bone," of which the first interpretation seems preferable because it better fits the preceding description of the blow "shearing the arm from the muscles. Thus, there is obviously a discrepancy between the weapon and the wound, since usually a spear should be a puncturing weapon, while in this case the wound it makes is clearly more extensive (cf. Saunders, 1999, 361; Saunders, 2003, 159–160; the scene is not considered by Hutchison and Hirthler, 2013. Though in most cases, the wound matches the weapon specified (cf. Robertson, 2002), this is not an isolated instance, and occasionally the depiction of both hit and wound are entirely plausible and realistic, but ultimately inconsistent and at odds with each other. Indeed, the description of the weapon(s) used by a warrior is sometimes in itself problematic, because Homer does not distinguish between the thrusting spear/lance and the throwing spear/javelin, sometimes does not specify the weapon at all, and occasionally has his heroes transition rapidly between weapons from killing to killing (cf. Saunders, 2003, 134–136; Saunders, 2004, 4). Such "wound–weapon mismatch" is the type of unrealistic description of battlefield trauma most easily explicable in light of Homeric composition: The poet drew on existing formulaic lines for both the mortal blow and the wound, but as he was composing, and improvising, in an oral performative context an incongruity between the formulae could slip in unnoticed (Saunders, 2003, 137). If this happened during an instance of dictation for the preservation of the poem, it was probably directly recorded in writing and thence transmitted. Additionally, while the loss of an arm is certainly a serious injury likely to ultimately result in death, it is probably not a cause for the "sudden death" described here (cf. Saunders, 2003, 132 who cites the more salient example of Amphiclus dying from a spear thrust to the thigh in *Il.* 16.313–316).

Another type of unrealism in Iliadic battle scenes are instances where the trauma depicted cannot be reconstructed satisfactorily, since it is not consistent with existing anatomical features. In the following example, the weapon is not specified, but in any case the wound is impossible (text: van Thiel, 2010; translation: Horn):

Il. *13.545–549: Ἀντίλοχος δὲ Θόωνα μεταστρεφθέντα δοκεύσας*
οὔτασ' ἐπαΐξας, ἀπὸ δὲ φλέβα πᾶσαν ἔκερσεν,
ἥ τ' ἀνὰ νῶτα θέουσα διαμπερὲς αὐχέν' ικάνει·
τὴν ἀπὸ πᾶσαν ἔκερσεν· ὃ δ' ὕπτιος ἐν κονίῃσι
κάππεσεν, ἄμφω χειρε φίλοις ἑτάροισι πετάσσας.

Antilochus, watching for Thoon turning his back,
leapt at him and wounded him, and severed the whole vein,
that, running all the way up along the back, reaches the neck:
this he severed entirely, and Thoon fell backwards into the dust,
stretching out both hands to his dear comrades.

As medically schooled commentators have pointed out, there simply is no such blood vessel running all the way up the back to the neck (cf., e.g., Friedrich, 2003, 34; Janko, 1994, 114; Saunders, 1999, 349–351; Saunders, 2003, 147–148). In this instance, the description seems quite plausible but is actually factually incorrect and must result from either faulty anatomical observation and/or interpretation or the imagination of the poet. In this category also fit cases, where the angle or trajectory of the weapon is at odds with anatomical realities or the relative positions of the combatants (for several examples, see the discussions in Saunders (1999), 346–362). Thus, this is another type of unrealistic and fantastical (albeit inconspicuously so) depiction of trauma, which has been termed "pseudo-realism" (Friedrich, 2003), since these wounds only appear to be lifelike and realistic but are problematic in detail upon closer examination.

The next passage is more imaginative and contains the strange and undignified image of the corpse of the slain warrior tumbling from his chariot and remaining upright stuck in the sand (text: van Thiel, 2010; translation: Horn):

Il. *5.580–589: Ἀντίλοχος δὲ Μύδωνα βάλ' ἡνίοχον θεράποντα*
ἐσθλὸν Ἀτυμνιάδην—ὃ δ' ὑπέστρεφε μώνυχας ιππους—
χερμαδίῳ ἀγκῶνα τυχὼν μέσον· ἐκ δ' ἄρα χειρῶν
ἡνία λεύκ' ἐλέφαντι χαμαι πέσον ἐν κονίῃσιν.
Ἀντίλοχος δ' ἄρ' ἐπαΐξας ξίφει ἤλασε κόρσην·
αὐτὰρ ὅ γ' ἀσθμαίνων εὐεργέος ἔκπεσε δίφρου
κύμβαχος ἐν κονίῃσιν ἐπι βρεχμόν τε και ὤμους.
δηθὰ μάλ' ἑστήκει, τύχε γάρ ῥ' ἀμάθοιο βαθείης,
ὄφρ' ιππω πλήξαντε χαμαι βάλον ἐν κονίῃσι,
τοὺς ιμασ' Ἀντίλοχος, μετὰ δὲ στρατὸν ἤλασ' Ἀχαιῶν.

Antilochus aimed at Mydon, the attendant and charioteer (sc. of Pylaemenes),
the noble son of Atymnius—he was just turning the single-hoofed horses around—
striking him with a boulder straight on the elbow; and from his hands
the reins, white with ivory, fell to the ground in the dust.
Then Antilochus pounced on him and drove his sword into his temple;
and gasping he fell from his well-wrought chariot
headlong into the dust on his forehead and shoulders.
For a very long time he stood there, for he had landed on deep sand,

until his two horses struck him down to the ground in the sand;
them whipped Antilochus and drove them to the army of the Achaeans.

This unique scene has recently been examined in detail by commentators, most recently by Kenneth Saunders with the eye of a modern physician as well as of a classical scholar. He concluded that despite prior attempts to explain the scene as an instance of cataleptic *rigor mortis*, the most likely solution to the reconstruction of the scene is that it is "fantastic and unrealistic," since "there is no known pathophysiological process which can render an entire corpse lastingly rigid at the moment of death" (Saunders, 2000, 32, referring to Friedrich, 2003, 9; Fenik, 1968, 64–65). This third type of unrealistic depiction of battlefield trauma has been characterized as "phantasma" since it introduces fantastical and even slightly grotesque elements into the narrative.

To illustrate the principle, another instance of a fantastical feature in death descriptions is the detail that a mortal blow to the head makes a warrior's eyes pop out and fall to the ground (*Il.* 13.615–617; 16.737–742). Lacking detailed medical knowledge, ancient readers accepted these descriptions as factual (cf. Janko, 1994, 122), but modern commentators noticed the problems with these scenes and have tried to construct them as cases of luxation of both eyeballs. However, they are probably better explained as cases of "melodramatic exaggeration" imagined by the poet (Saunders, 1999, 351–352): In these instances of intense fighting, the unrealistic element might be a fictive result meant to highlight the devastating force of the blow. In this respect, "phantasmata" are likely similar to the depiction of battle in modern cinema: There even ostensibly realistic, historical movies may contain particularly gruesome, imaginative, and medically incorrect scenes to emphasize the intensity and relentlessness of combat.

In conclusion, Classical scholars have not focused their attention on the content and the reconstruction of anatomical detail in Iliadic battle descriptions. Not being medical professionals and lacking the necessary anatomical and pathological knowledge, their main objective has been the examination of the poem's formulaic language, its conventions and patterns as well as their meaning in the context of heroic epic. Thus the extent and elements of stylization of battle scenes (frequent, sudden death; few wounds; no nonlethal mutilations) in the narrative of the *Iliad* have been explained by their purpose of highlighting the excellence and fortitude of individual Homeric heroes.

However, beyond this literary critical focus, Classicists have also been grateful for, and availed themselves of, the input on the Homeric poems provided by scholars from different background. Thus, philologists have been interested in the findings of archeologists and historians regarding the historicity of Homeric epic, who have determined that the material and sociocultural world depicted in the Homeric poems is an amalgam of features from several historical periods mixed with some ahistorical and fantastical elements. Based on the studies of scholars with medical expertise, Classicists are not surprised to encounter a similar state of affairs regarding battle scenes. Most descriptions are clearly pathophysiologically accurate and realistic, but this obviously does not apply to all of them. Kenneth Saunders has calculated that of the 64 wounds described in *Iliad* 13–16, a total of 14 (approx. 21%) require further debate (Saunders, 1999, 346), as they contain some problematic or unrealistic features in varying degrees. Saunders explained these findings as the result of poetic conventions and formulaic slips as well as by outlining the limited anatomical knowledge we can expect on the part of an Archaic Greek poet (and his audience, for whom the epics were tales from the past). It is undoubtedly correct that the form of Iliadic battle scenes is the product of traditional diction and poetic stylization, and the studies of oral poetic technique can account for some of the details in need of explanation, such as wound–weapon mismatches. Yet Saunder's comment of Homer's limited medical knowledge is certainly true (after all, even medical professionals then had only a very limited grasp on the workings of the human body from our modern point of view), but also slightly beside the point from a Classicist's perspective: Just as we can assume for modern film directors, medical realism was also not the main objective of the poet as he composed his battle scenes (and neither was his audience much interested in receiving an anatomy lesson). Rather, Homer was primarily concerned with providing his listeners with an engrossing narrative and cinematic entertainment, of which his riveting depictions of war are just one component (also cf. Gazis, 2023). In this, he was certainly very successful and readers of all professions and backgrounds are still fascinated and enthralled by his poems.

References

Albracht, F. (2005). *Battle and battle description in Homer. A contribution to the history of war*. Duckworth.

Apostolakis, E., Apostolaki, G., Apostolaki, M., & Chorti, M. (2010). The reported thoracic injuries in Homer's Iliad. *Journal of Cardiothoracic Surgery, 5*, 114.

Armstrong, C. B. (1969). The casualty lists in the Trojan War. *Greece & Rome, 16*, 30–31.

Fenik, B. (1968). *Typical battle scenes in the Iliad: Studies in the narrative techniques of Homeric battle description*. Franz Steiner Verlag.

Friedrich, W.-H. (2003). *Wounding and Death in the Iliad: Homeric Techniques of Description*. London: Duckworth.

Frölich, H. (1879). *Die Militärmedicin Homers*. Verlag von Ferdinand Enke.

Gazis, G. A. (2023). Homer and the art of cinematic warfare. In A. Capra, & L. Floridi (Eds.). *Intervisuality. New approaches to Greek Literature* (pp. 81–101). De Gruyter.

Griffin, J. (1980). *Homer on life and death*. Clarendon Press.

Hellmann, O. (2000). *Die Schlachtszenen der Ilias: Das Bild des Dichters vom Kampf in der Heroenzeit Hermes Einzelschriften*. Franz Steiner Verlag.

Horn, F. (2014). *Held und Heldentum bei Homer: Das homerische Heldenkonzept und seine poetische Verwendung Classica Monacensia*. Narr Verlag.

Horn, F. (2018). Dying is hard to describe: Metonymies and metaphors of death in the Iliad. *Classical Quarterly, 68*, 359–383.

Hutchison, R. L., & Hirthler, M. A. (2013). Upper extremity injuries in Homer's Iliad. *History of Hand Surgery, 38*, 1790–1793.

Janko, R. (1994). *The Iliad: A commentary. Volume IV: Books 13–16 IV*. Cambridge University Press.

Kayhanian, S., & Machado, R. J. (2020). Head injuries in Homer's Iliad. *World Neurosurgery, 143*, 33–37.

Kömürcu, E., Tok, F., Şimşek, A., & Özçakar, L. (2014). Musculoskeletal injuries in Homer's Iliad: The War of Troy Revisited. *American Journal of Physical Medicine & Rehabilitation, 93*, 335–341.

Körner, O. (1929). *Die ärztlichen Kenntnisse in Ilias und Odyssee*. Verlag von J. F. Bergmann.

Koutserimpas, C., & Samonis, G. (2016). Trauma cases in Homer's Odyssey. *Journal of Wound Care, 25*, 553–554.

Lane Fox, R. (2020). *The invention of medicine: From Homer to Hippocrates*. Allen Lane.

Lipshutz, H. (1963). Hand, arm, and shoulder trauma in the Iliad. *Surgery, 54*, 833–836.

Majno, G. (1975). *The healing hand: Man and wound in the ancient world*. Harvard University Press.

Marg, W. (1976). Kampf und Tod in der Ilias. *Würzburger Jahrbücher für die Altertumswissenschaft n.s. 2*, 7–19.

Morrison, J. V. (1999). Homeric darkness: Patterns and manipulation of death scenes in the ›Iliad‹. *Hermes, 127*, 129–144.

Mylonas, A. I., Tzerbos, F. H., Eftychiadis, A. C., & Papadopoulou, E. C. (2008). Craniomaxillofacial injuries in Homer's Iliad. *Journal of Cranio-Maxillofacial Surgery, 36*, 1–7.

Neal, T. (2006). *The wounded hero: Non-fatal injury and bloodspill in Homer's Iliad*. Peter Lang.

Niens, C. (1987). *Struktur und Dynamik in den Kampfszenen der Ilias*. Julius Groos Verlag.

Nomikos, N. (2018). Injuries in the Greek epics of Homer. *Chinese Journal of Traumatology, 21*, 109–112.

Ralli, I., Stathopoulos, P., Mourouzis, K., Piagkou, M., & Rallis, G. (2015). Facial trauma in the Trojan War. *Oral and Maxillofacial Surgery, 19*, 191–194.

Robertson, C. (2002). Wounds and wounding in the Iliad. *Ancient History Bulletin, 16*, 103–110.

Salazar, C. F. (2000). *The treatment of war wounds in Graeco-Roman antiquity studies in ancient medicine*. Brill.

Santos, G. H. (2000). Chest trauma during the battle of troy: Ancient warfare and chest trauma. *The Annals of Thoracic Surgery, 69*, 1285–1287.

Sapounakis, C., Rallis, G., Mourouzis, C., Konsolaki, E., & Tesseromatis, C. (2007). Injuries to the head and neck in Homer's Iliad. *British Journal of Oral and Maxillofacial Surgery, 45*, 112–115.

Saunders, K. B. (1999). The wounds in Iliad 13-16. *Classical Quarterly, 49*, 345–363.

Saunders, K. B. (2000). A note on the strange death of mydon in Iliad 5. *Symbolae Osloenses, 75*, 24–33.

Saunders, K. B. (2003). *Appendix to W. H. Friedrich, Wounding and death in the Iliad: Homeric techniques of description*. Duckworth, 131–167.

Saunders, K. B. (2004). Frölich's table of homeric wounds. *Classical Quarterly, 54*, 1–17.

Stoevesandt, M. (2004). *Feinde —Gegner —Opfer. Zur Darstellung der Troianer in den Kampfszenen der Ilias* Schweizerische Beiträge zur Altertumswissenschaft. Schwabe.

Strasburger, G. (1954). *Die kleinen Kämpfer der Ilias*. PhD thesis, Frankfurt am Main.

van Thiel, H. (2010). *Homeri Ilias* (2nd ed.). Georg Olms Verlag.

Thompson, A. R. (1952). Homer as a surgical anatomist. *Proceedings of the Royal Society of Medicine, 45*, 765–767.

van Wees, H. (1986). Leaders of men? Military organisation in the Iliad. *Classical Quarterly, 36*, 285–303.

van Wees, H. (1988). Kings in combat: Battles and heroes in the Iliad. *Classical Quarterly, 38*, 1–24.

van Wees, H. (1994). The Homeric way of war: The Iliad and the hoplite phalanx. *Greece & Rome, 43*, 1-18–131-155.

Section 3

Neurology, toxicology, infectious disease and genetics in homer

Neurotrauma and other neurology in Homer

15

Eric L. Altschuler[1,2], *and Anusha Lekshminarayanan*[3,4]
[1]Essen Health Care, Bronx, NY, United States, [2]Kessler Institute for Rehabilitation, West Orange, NJ, United States, [3]Department of Rehabilitation Medicine, New York Medical College, Valhalla, NY, United States, [4]Department of Physical Medicine and Rehabilitation, Westchester Medical Center, Valhalla, NY, United States

The many cases of head and neck injuries and neurotrauma in the *Iliad* are listed in Tables 13.1 and 13.2. See also Mylonas et al. (2008). Here we discuss one remarkable case in detail. We also mention a case of non-neurotrauma neurologic disease in Homer and a case of neurotrauma/neurologic disease not involving a head or neck injury:

First description of a brachial plexus injury

The *Iliad* appears to contain the first description of brachial plexus injury (Aydn et al., 2004; Basma et al., 2022; Hutchison & Hirthler, 2013; Sahlas, 2001; Thompson, 1952).

In Book 8 of the *Iliad* (lines 320–329) Trojan (crown) prince Hector injures Greek archer Teucer with the throw of a stone that lands on Teucer's arm:

Translation from Homer (1924). This work is licensed under a Creative Commons Attribution-ShareAlike 3.0 United States License.

[320] And himself Hector leapt to the ground from his gleaming car crying a terrible cry, and seizing a stone in his hand made right at Teucer, and his heart bade him smite him. Now Teucer had drawn forth from the quiver a bitter arrow, and laid it upon the string, but even as he was drawing it back Hector of the flashing helm [325] smote him beside the shoulder where the collar-bone parts the neck and the breast, where is the deadliest spot; even there as he aimed eagerly against him he smote him with the jagged stone, and he brake the bow-string; but his hand grew numb at the wrist, and he sank upon his knees and thus abode, and the bow fell from his hand.

Fig. 15.1 demonstrates the brachial plexus injury.

We must wonder how Homer knew this trauma would cause a brachial plexus injury!

Elpenor syndrome

Homer was the first (Logre, 1961) to describe the sleep parasomnia CONFUSIONAL AROUSAL (da Silva Macedo et al., 2023; Singh et al., 2018): Individuals with the condition are confused and disoriented when attempting to fall asleep, typically after

Medicine in Homer. DOI: https://doi.org/10.1016/B978-0-443-32978-4.00004-3

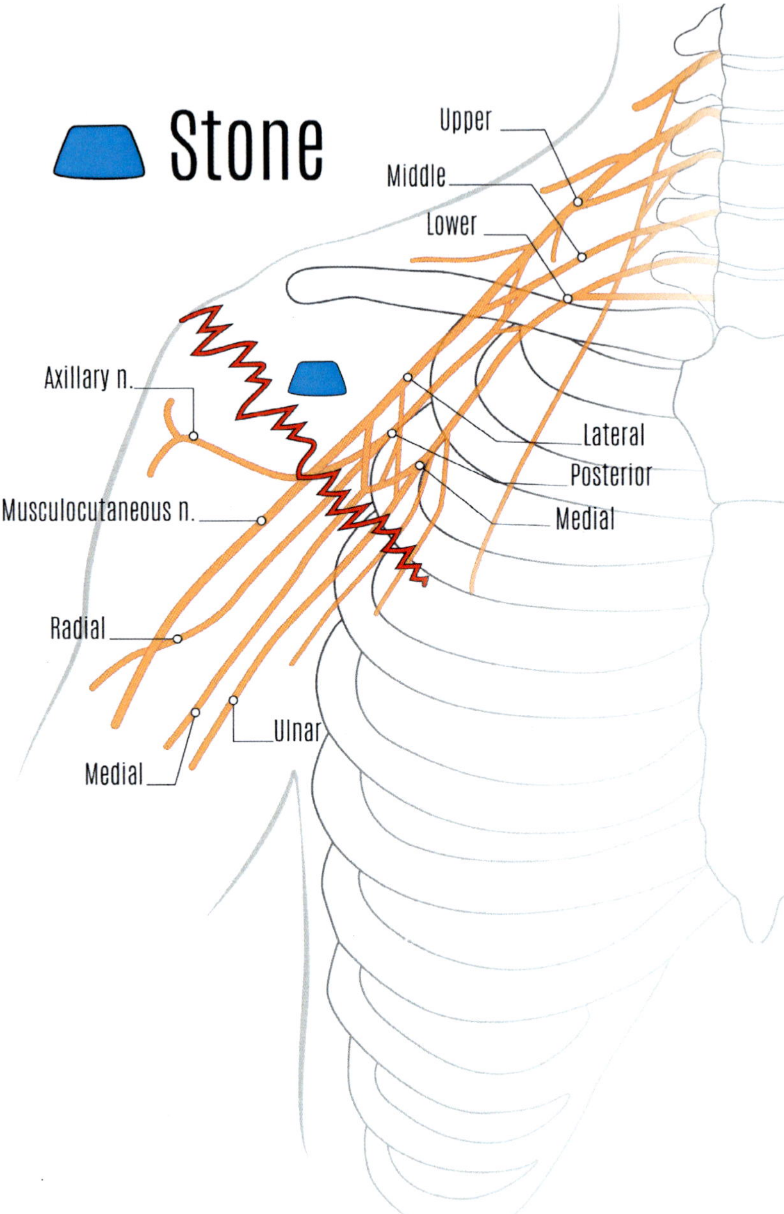

Figure 15.1 Brachial plexus injury. The injury to Teucer's brachial plexus caused by the stone thrown by Hector is shown.
Source: Image by Mileidy Fernandez after an image drawn by Anusha Lekshminarayanan.

excessive drinking of alcohol or use of sedative medications and may experience half-automatic movements that can lead to accidents.

The conditional is also known as Elpenor syndrome (Logre, 1961; Olry and Haines, 2006) named for the youngest of Odysseus' colleagues in the *Odyssey* journeying back with him from Troy to Ithaca who died from a broken neck (Walshe, 1997) after awakening from a fitful attempt at sleep while drunk: *Odyssey*, Book 10, lines 552–560[1]. (Translation from Homer (1919). This work is licensed under a Creative Commons Attribution-ShareAlike 3.0 United States License.)

There was one, Elpenor, the youngest of all, not over valiant in war nor sound of understanding, who had laid him down apart from his comrades in the sacred house of Circe, [555] seeking the cool air, for he was heavy with wine. He heard the noise and the bustle of his comrades as they moved about, and suddenly sprang up, and forgot to go to the long ladder that he might come down again, but fell headlong from the roof, and his neck [560] was broken away from the spine, and his spirit went down to the house of Hades.

Decerebrate positioning

Kayhanian and Machado (2020) present a fascinating analysis of the injury Ilioneus sustains due to a spearing by Peneleos (*Iliad*, Book 14, lines 493–496) which is immediately followed by Ilioneus' beheading by Peneleos (using a sword, lines 497–499 see also Case 26 in Table 13.2):

Greek and translations from Homer (1924). This work is licensed under a Creative Commons Attribution-ShareAlike 3.0 United States License.

Iliad, Book 14, lines 493–496

τὸν τόθ᾽ ὑπ᾽ ὀφρύος οὖτα κατ᾽ ὀφθαλμοῖο [eyes] θέμεθλα,
ἐκ δ᾽ ὦσε γλήνην: δόρυ δ᾽ ὀφθαλμοῖο διὰ πρὸ
495 καὶ διὰ ἰνίου [inion] ἦλθεν, ὃ δ᾽ ἕζετο **χεῖρε πετάσσας**
ἄμφω: Πηνέλεως δὲ ἐρυσσάμενος ξίφος ὀξὺ

Him [Ilioneus (a Trojan)] then did Peneleos [a Greek] smite beneath the brow at the roots of the **eyes**, and drave out the eyeball, and the shaft went clean through the eye [495] and through the nape of the neck, and he sank down **stretching out both his hands**. [497 But Peneleos drawing his sharp sword let drive full upon his neck, and smote off to the ground the head with the helmet, and still the mighty spear stood in the eye]

First, Kayhanian and Machado (2020) agree with Saunders' (Saunders, 1999) analysis that the path of the spear entering in the eye (line 493) and exiting at the INION (a prominent notch in the back of the scalp that you can feel on yourself) must have passed through the midbrain (see Fig. 15.2).

Next, given presumed damage to the midbrain and Homer's use of the phrase **χεῖρε πετάσσας** (**stretching out both his hands**) Kayhanian and Machado claim that

[1] Homer's knowledge of the specifics and time course of the syndrome suggest he likely observed this on his voyage to or from the Troad.

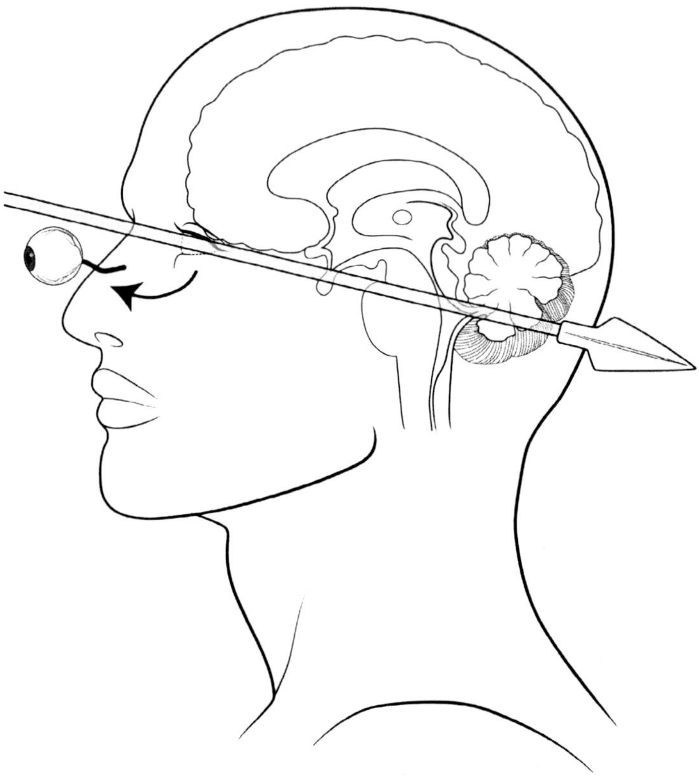

Figure 15.2 Midbrain injury. Decerebrate posturing as a result of midbrain injury with spear (entering through the eye, passing through the midbrain, and exiting through the nape of the neck).
Source: Image by Mileidy Fernandez after an image drawn by Anusha Lekshminarayanan.

Ilioneus is lying in a **DECEREBRATE** position (see Fig. 15.3). Decerebrate positioning is a sign of severe brain damage (Abnormal Posturing, 2024) worse still than DECORTICATE positioning (arms crossed over the chest) and a lesion to the midbrain can cause decerebrate positioning ((Abnormal Posturing, 2024) and refs. therein).

To demonstrate that by **χεῖρε πετάσσας** Homer indeed meant **stretching out both hands** Kayhanian and Machado (2020) perform an analysis akin to what is known in legal research as "Sherpardzing" (Shepard's Citations, 2024) the phrase (case) which is named after a legal service begun by Frank Shepard (1848–1900) in 1873, when Shepard began publishing lists of citations to a given case from different jurisdictions. Analogously, Kayhanian and Machado look at all of Homer's other uses of **πετάσσας** (the verb being inflected as needed) and they see that in each of a number of different contexts the word indeed means "stretching out":

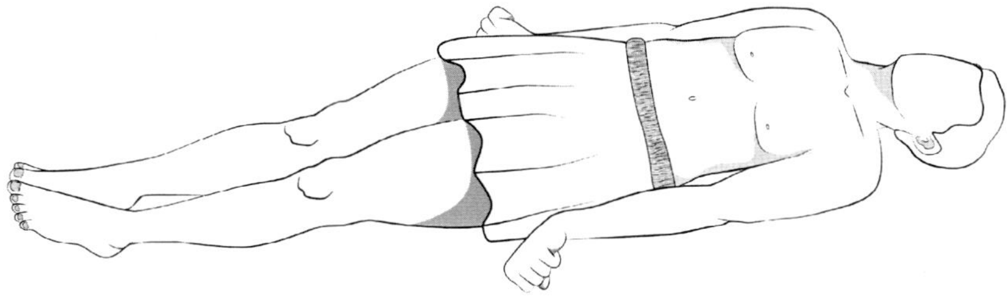

Figure 15.3 Decerebrate position. The patient is lying in a decerebrate position with his arm and hands at his sides and extended.
Source: Figure by Mileidy Fernandez after Copyright free https://commons.wikimedia.org/wiki/File:Decerebrate.jpg.

For example, to spread out a sail:
Iliad, Book 1, line 480
οἳ δ᾽ ἱστὸν στήσαντ᾽ ἀνά θ᾽ **ἱστία λευκὰ πέτασσαν**,
and they set up the mast and **spreadthe white sail**.
Odyssey, Book 4, line 783
πάντα κατὰ μοῖραν, ἀνά θ᾽ **ἱστία λευκὰπέτασσαν**:
all in due order, and **spreadthe white sail**.
Odyssey, Book 5, line 269
γηθόσυνος δ᾽ οὔρῳ **πέτασ᾽ ἱστία** δῖος Ὀδυσσεύς.
Gladly then did goodly Odysseus **spread his sail** to the breeze
Or a person stretching out his hands,
Iliad, Book 4, line 523
κάππεσεν ἄμφω **χεῖρε** φίλοις ἑτάροισι **πετάσσας**
and he fell backward [in the dust] and **stretched out both his hands** to his dear comrades
Iliad, Book 13, line 549
κάππεσεν, ἄμφω **χεῖρε** φίλοις ἑτάροισι **πετάσσας**.
[and Thoön] fell on his back in the dust, **stretching out both his hands** to his dear comrades.
Odyssey, Book 9, line 417
αὐτὸς δ᾽ εἰνὶ θύρῃσι καθέζετο **χεῖρε πετάσσας**,
and himself sat in the doorway with **arms outstretched**
or a cloth being stretched out
Iliad, Book 8, line 441
ἅρματα δ᾽ ἂμ βωμοῖσι τίθει κατὰ λῖτα **πετάσσας**:
[unyoked his horses] and set the car upon a stand, and **spread** thereover a cloth
or clothing being spread (stretched) out.
Odyssey, Book 6, line 94
ἑξείης **πέτασαν** παρὰ θῖν᾽ ἁλός, ἧχι μάλιστα
they **spread them out** in rows on the shore of the sea
Finally, getting back to a person stretching out arms:

Odysseus' posture when diving into the sea
Odyssey, Book 5, line 374
αὐτὸς δὲ πρηνὴς ἁλὶ κάππεσε, **χεῖρε πετάσσας**,
and flung himself headlong into the sea with **hands outstretched**
to express supplication,
Iliad, Book 21, line 115
ἔγχος μέν ῥ' ἀφέηκεν, ὃ δ' ἕζετο **χεῖρε πετάσσας**
The spear he let go, but crouched with **both hands outstretched**.
affection
Odyssey, Book 24, line 397
ὣς ἄρ ἔφη, Δολίος δ' ἰθὺς κίε **χεῖρε πετάσσας**
So he spoke, and Dolius ran straight toward him with **both hands outstretched**
or a plan (plea) for help.
Iliad, Book 13, line 821
ὣς ἄρα οἱ εἰπόντι **ἐπέπτατο** δεξιὸς **ὄρνις**
Even as he thus spake, there **flew forth a bird** upon the right hand,

This analysis confirms that after the spear blow from Peneleos Ilioneus was indeed lying in a decerebrate position. Homer's description of decerebrate positioning is particularly impressive because the modern description was not given until 1898 (Sherrington, 1898) by Sir Charles Scott Sherrington (1857–1952) the 1932 Nobel Laureate in Medicine or Physiology (with Edgar Douglas Adrian) "for their discoveries regarding the functions of neurons." Furthermore, the time course of Ilioneus' injuries: spear blow entering in the eye, going through the midbrain then exiting at the inion; decerebrate positioning; sword blow to the neck causing death makes it likely that Homer observed these events rather than having to rely on some other observer reporting a Nobel Prize level discovery to him.

References

Abnormal Posturing (2024). https://en.wikipedia.org/wiki/Abnormal_posturing.

Aydn, A., Tümerdem, B., Yazar, M., & Erer, M. (2004). Three-thousand-year-old written reference to a description of what might be the earliest brachial plexus injuries in the Iliad of Homer. *Plastic and Reconstructive Surgery, 114*(5), 1352–1353.

Basma, J., Gienapp, A. J., Arnautovic, K. I., & Konofaos, P. (2022). First documented clinical account of brachial plexus palsy from the 12th century. *Journal of Neurosurgery, 136*(4), 1179–1185. https://doi.org/10.3171/2021.4.JNS21218.

da Silva Macedo, L. J., Alves, A. O., Mazza, G. S., Pinto, A. S. B., Coelho, F. M. S., & Oliveira, G. D. P. (2023). Sleeping and dreaming in Greek mythology. *Sleep Medicine, 101*, 178–182. https://doi.org/10.1016/j.sleep.2022.10.020.

Homer (1919). *The Odyssey*. Harvard University Press.

Homer (1924). *The Iliad*. Harvard University Press.

Hutchison, R. L., & Hirthler, M. A. (2013). Upper extremity injuries in Homer's Iliad. *The Journal of Hand Surgery, 38*(9), 1790–1793. https://doi.org/10.1016/j.jhsa.2013.06.022.

Kayhanian, S., & Machado, R. J. (2020). Head Injuries in Homer's Iliad. *World neurosurgery, 143*, 33–37. https://doi.org/10.1016/j.wneu.2020.07.037.

Logre, B. J. (1961). *Psychiatrie clinique. PsychiatPresses universitaires de France Ven-dome.* Paris PsychiatPresses universitaires de France Ven-dome.

Mylonas, A. I., Tzerbos, F. H., Eftychiadis, A. C., & Papadopoulou, E. C. (2008). Cranio-maxillofacial injuries in Homer's Iliad. *Journal of Cranio-Maxillo-Facial Surgery, 36*(1), 1–7. https://doi.org/10.1016/j.jcms.2007.07.010.

Olry, R., & Haines, D. E. (2006). NEUROwords: Elpenor's syndrome: The link between one of Ulysses' companions and the tenth president of the Third French Republic. *Journal of the History of the Neurosciences, 15*(2), 159–161.

Sahlas, D. J. (2001). Functional neuroanatomy in the pre-Hippocratic era: Observations from the Iliad of Homer. *Neurosurgery, 48*(6), 1352–1357.

Saunders, K. B. (1999). The wounds in "Iliad" 13-16. *Classical Quarterly, 49*(2), 345–363.

Shepard's Citations (2024). https://en.wikipedia.org/wiki/Shepard%27s_Citations.

Sherrington, C. S. (1898). Decerebrate rigidity, and reflex coordination of movements. *The Journal of Physiology, 22*(4), 319–332.

Singh, S., Kaur, H., Singh, S., & Khawaja, I. (2018). Parasomnias: A comprehensive review. *Cureus, 10*(12). https://doi.org/10.7759/cureus.3807.

Thompson, A. R. (1952). Homer as a surgical anatomist. *Proceedings of the Royal Society of Medicine, 45*(11), 765–767.

Walshe, T. M. (1997). Neurological concepts in archaic Greece: What did Homer know? *Journal of the History of the Neurosciences, 6*(1), 72–81.

A case of antisocial personality disorder in Homer

16

Eric L. Altschuler[1,2]
[1]Essen Health Care, Bronx, NY, United States, [2]Kessler Institute for Rehabilitation, West Orange, NJ, United States

As described in the prior chapter (Chapter 15) most of the neurology in Homer is neurotrauma. This contrasts with Shakespeare in whose works neurologic and neuropsychiatric disease is not uncommon: e.g., epilepsy in *Julius Caesar* (Act 1, Scene 2), posttraumatic stress disorder (PTSD) *Henry IV, Part 1* (Act 2, Scene 3 lines 39–67) and *Romeo & Juliet* (Act 1, Scene 4, lines 81–85) and hepatic encephalopathy in Sir Andrew Aguecheek from *Twelfth Night* (Summerskill, 1955). (Both Homer and Shakespeare describe disorders associated with sleep such as confusional arousal also known as Elpenor Syndrome after the character in the *Odyssey* (see Chapter 15), obstructive sleep apnea in Shakespeare's Falstaff (see *Henry IV, Part 1*, Act 2, Scene 4, 470–473) and sleep walking and talking (*Macbeth*, Act 5, Scene 1).

One psychiatric disease which is in Homer is antisocial personality disorder (ASPD). In medical school, we learned that psychiatric disease is diagnosed by means of an SCID—structured clinical interview for diagnosis. If there is no general medical problem, then via the interview a diagnosis can be obtained. I realized this can be used to diagnose literary characters as great writers rule out general medical disease if there is sufficient text. The oldest description of ASPD is the case of Samson the Bible from the Book of Judges, Chapters 13–16 (Altschuler et al., 2001; Altschuler, 2017). See Box 16.1.

What about Heracles? Unfortunately, unlike the biblical story of Samson there is not one old, long coherent text, however, I think what information we do have is sufficient to secure the diagnosis (Altschuler, 2017):

Medicine in Homer. DOI: https://doi.org/10.1016/B978-0-443-32978-4.00009-2

The earliest literary (or artistic) mentions of Heracles can be found in Homer's *Iliad* and *Odyssey*.[1] (The most substantial and longest ancient text on Heracles comes from the much later Latin poet Ovid (43 BC–17 AD) in Book 9 of his *Metamorphoses*.) Book 5, lines 648–651 of The *Iliad* describe how, in a past time, Heracles destroyed Troy after he was treated poorly by the Trojan King Laomedon (father of the Trojan King Priam of Homeric fame). Apparently, Heracles was provoked, but his response—destroying a city—seems a bit extreme. Indeed, the thoroughness of the destruction is emphasized in Book 14, line 251. Heracles is featured in two places in the *Odyssey*. In Book 11, Odysseus visits Hades to get advice, counsel, and information from various "shades" and "spirits" there. He encounters Heracles near the very end of the book (lines 599–628). Heracles does not act in a particularly antisocial way in this Book, but Homer notes that the other spirits were frightened of Heracles, who was carrying a bow and arrow. Odysseus' hackles were raised as well. At the very least this shows that Heracles was a frightening and potentially violent individual, though Heracles commiserates with Odysseus, comparing Odysseus' wanderings to his own labors. In Book 21 (lines 24–29), we find the strongest evidence that Heracles may have had ASPD. Here he kills Iphitos in order to steal and keep Iphitos' horses. There is no sign that Heracles shows any remorse for his act. Today, in many jurisdictions, this would carry a sentence of death because, while committing murder, Heracles also committed another felony—robbery. Samson committed this same heinous dual crime (see Box 16.1, criteria 7). Thus in Homer, we see that Heracles certainly meets the ASPD criteria for aggressiveness, as he is repeatedly involved in fights and physical

[1] When reading the Iliad and Odyssey—the oldest written documents of a culture—and paying attention to the facts, or what has been written as being already known, the reader can catch a glimpse of what was known in the time before the texts were written. Along with Heracles and his labors—which were clearly well known to Homer's readers—here are some other examples of references to Greek pre-Epic culture: The women warriors of Amazonia are mentioned in Iliad Book 3. The Iliad (Book 7, line 468 and Book 23, line 746) mentions Jason without need for further elaboration. This indicates that Jason's quest as captain of the ship Argo leading heroes including Heracles to the Black Sea coast to retrieve the Golden Fleece was a story well-known already to Homer's readers. The Odyssey Book 4, line 130 mentions Alkandre's travels to Egypt. Herodotus in his Histories appreciates the travels of this long-famed traveler. In Book 8, line 224 the bowman Eurytos is mentioned. Eurytos was so proud of his skill in archery, he challenged Apollo to a contest. Apollo then killed him in anger. Book 10 of the Odyssey line 496 mentions the blind Thebean seer Theiresias in a manner indicating that Theirsias was a figure already known to readers of the Odyssey. An interesting recurring story in The Iliad (first mentioned in Book 10, line 17) and the Odyssey is that of Neleus, Nestor's father and his predecessor as king of Pylos. As it is, Nestor is the wise old man of the Iliad and Odyssey. Nestor seems at least two generations older than other characters in the epics. Neleus is mentioned in the Odyssey Book 3, lines 406–409, Book 11, line 254 and lines 282–286 which describe how Neleus wooed and married the lovely Khloris and the birth of Nestor. The exploits of Neleus are also mentioned in Book 15 lines 229–240. So Neleus is known to be from the day both in mythologic time and "TV time." Nestor, interestingly, may not have been as well-known to Homer's thus requiring his long "origin story" in Book 11 of the Iliad lines 690–760. Attempting to partially console Priam when he comes to retrieve Hector's body from the Greek's camp at their ships Achilles (Iliad Book 24, lines 602–617) recounts the story of Niobe whose six sons were killed by arrows shot by Apollo and her six daughters similarly by arrows shot by Artemis whom Niobe had angered by comparing her six daughters to Artemis' two. Interestingly, in Achilles' telling of the tale Niobe's people were turned to stone by Zeus, a detail as Murray notes (Homer, 1999, p. 609) which is unknown in other versions of the story.

Box 16.1 Diagnosing Samson, Son of Manoah, with ASPD.

Samson, son of Manoah, meets six of the seven criteria for ASPD (Judges, Chapters 13–16; Altschuler et al., 2001). But in a clever way, the writer is also letting us know that he knows the seventh criterion as well.

Most of the criteria Samson meets more than once, and the author reinforces this point again and again, just to make sure we get it. Also, criteria are often met in a clear way. For example, one criterion for ASPD is reckless disregard of safety of self. Well, Samson took on 1000 men all by him-self in one encounter. Disregard of safety of self might be taking on two or three men at once. Reckless disregard of safety of self might be taking on ten men at once. Samson takes this to a hyperbolic extreme in taking on a thousand at once! The Judges tale demonstrates point by diagnostic point:

1. Failure to conform to social norms with respect to lawful behavior: The Philistines tried to arrest Samson after he burned the Philistine fields (Judges Chapter 15: verse 5) and went to Gaza (16:1).
2. Deceitfulness, as indicated by repeated lying: Samson did not tell his parents that he had killed a lion. Furthermore, he proffered honey for his parents to eat, but did not tell them it had come from the carcass of a lion (14:9) and thus caused them to violate their dietary laws. Samson also demonstrated deceitfulness or cunning in using a riddle that it would be impossible for the interlocutors to answer (Judges 14:12–14).
3. Impulsivity: His burning of the Philistine fields (15:5).
4. Irritability and aggressiveness: This is indicated by his repeated involvement in physical fights, almost none of which are planned or organized beforehand. They're basically hotheaded brawls.
5. Reckless disregard for safety of self or others: Samson is reported to have taken on and killed 1000 Philistines single-handedly (15:15). To translate this mathematically: if one individual takes on one thousand, this one individual would have to believe in having a probability of greater than 99.93% of beating each opponent to even assume to have a 50% chance of surviving the fight. Similarly, telling Delilah the secret to his strength (16:17)—even after she had obviously attempted three times previously to get this secret—can also be considered reckless disregard for safety of self.
6. Lack of remorse: He gloated (15:16) after killing 1000 men.
7. The only criterion for ASPD that Samson does not meet is failure to meet one's financial obligations. However, the author of the story (with a wink and a nod) betrays knowledge of this criterion as well when he has Samson repay his lost bet only by killing 30 men and using the stolen dead men's garments to pay up (Judges 14:19). So, the author knows all seven criteria for ASPD and is clearly letting us know that he knows.

In addition to these adult criteria, to diagnose ASPD an individual must meet criteria when he was a youth. Unbelievably—and again, as if the writer had a 3,000-year advance copy of the DSM—the author lays them out as if he were a textbook example. Samson committed many of the actions in the criteria for conduct disorder—fire setting, cruelty to small animals (15:5), bullying, initiating physical fights, using a weapon (like the donkey's jawbone) (15:15), and stealing from a victim (14:19). If the preliminary conduct disorder did not start when Samson was younger than 15 years, he was still quite young (14:1–6).

Author's own work

acts of violence. Heracles, too, can show lack of remorse. At times he also shows disregard for the safety of others.

Another ingenious medical point about Heracles was made by Branimir Čatipović in a short letter (Catipović, 2004) in the *New England Journal of Medicine*. A protein called myostatin was found some years previously, which, when mutated, causes increased muscle mass. This mutation was found first in mice, then in cattle, and then in a case report (Schuelke et al., 2004) of a human child that Čatipović commented on. "Myo" is the Greek word for muscle and "statin" is the Latin word for stop; this protein puts the brakes on muscle growth, lest it get too big and bulky for the rest of the body to support. The mice, cows, and the child had much greater muscle bulk than is considered within normal range. For example, the child had substantial muscle bulk observable at birth, and at age 4.5 years could hold two 3 kg weights out horizontally. Čatipović made the ingenious suggestion that this child may not have been the first case of myostatin deficiency in humans, but that Heracles was! Čatipović noted Heracles' great strength from birth as evidenced by his killing of a snake with each hand while an infant. Also, as Čatipović pointed out, Heracles' strength was never attributed to an exercise or dietary regimen. One wonders if there could have been a real person who had myostatin deficiency that Heracles was modeled after.

I noticed (Altschuler, 2001) an interesting medical footnote about Heracles seeing an exhibit, The Golden Deer of Eurasia: Scythian and Sarmatian Treasures from the Russian Steppes, at the Metropolitan Museum of Art (Aruz *et al.*, 2000). This exhibit had remarkable pieces from the Russian Steppes. One particular piece, a vessel or a cup featured a relief frieze that showed a timeless exemplar of medical compassion and care: One man tending the wounds of another. Additional figures on the cup told the story of the founding myth of Scythia as recounted in Book 4.8–10 of Herodotus' *Histories*: Heracles had been to the land of Scythia where a strange maiden/serpent consort of his bore him three children. He told the mother-to-be that when the children grew older to have them each try to string a bow that Heracles left behind. The one who did would be King. Neither of the two older brothers, Agathyrsus or Gelonus, could string the bow. (The injuries they sustained while trying are shown on the vessel.) But the youngest son, Scythes, could string the bow, and the kingdom was thus founded. This footnote turns out to have a footnote of its own. Using the ability to string a bow as a test of kingship is not confined to Scythes and the Scythian founding myth. Homer famously uses it in Book 21 of the *Odyssey*! Potential suitors for Odysseus' wife Penelope tried to string Odysseus' bow. (Of course, only Odysseus in disguise could do so.) Clearly there is a Freudian component to the task, which could have led to independent discovery and use of this task by the authors of the Scythian founding myth and Homer in the *Odyssey*. However, I find it more likely that the author of one myth borrowed the idea from the other. Or might somehow the same author have composed both stories?

References

Altschuler, E. L., Haroun, A., Ho, B., & Weimer, A. (2001). Did Samson have antisocial personality disorder? *Archives of General Psychiatry, 58*(2), 202–203.

Altschuler, E. L. (2001). Art. *The Lancet, 357.*

Altschuler, E. L. (2017). *A great literature guide to the DSM-5.* Oxford.

Aruz, J., Farkas, A., Alekseev, A., & Korolkova, E. The golden deer of Eurasia: Scythian and Sarmatian treasures from the Russian steppes. The State Hermitage, Saint Petersburg, and the Archaeological Museum. In: J. Aruz, A. Farkas, A. Alekseev, & E. Korolkova (Eds.), *Metropolitan Museum of Art.* Yale University Press.

Catipović, B. (2004). Myostatin mutation associated with gross muscle hypertrophy in a child. *The New England Journal of Medicine, 351*(10).

Schuelke, M., Wagner, K. R., Stolz, L. E., Hübner, C., Riebel, T., Kömen, W., Braun, T., Tobin, J. F., & Lee, S.-J. (2004). Myostatin mutation associated with gross muscle hypertrophy in a child. *The New England Journal of Medicine, 350*(26), 2682–2688.

Summerskill, W. H. (1955). Aguecheek's disease. *Lancet (London, England), 269*(6884).

Homer's pharmacology, toxicology, and epidemics through the lens of current scientific evidence

17

Andreas Plaitakis[1,2]
[1]University of Crete, Heraklion, Greece, [2]Icahn School of Medicine at Mount Sinai, New York, NY, United States

Introduction

The *Iliad* and *Odyssey* are masterworks of European literature attributed to Homer that date back to about three millennia ago. Owing to their ingenuity and enduring artistic value, the poems are still widely read. In addition to glorifying heroism and honor, the epics are remarkable for their seminal content which covers many disciplines, including medicine, psychology, pharmacology, social customs, ethics, agriculture, farming, seamanship, and astronomy. Homer also describes in detail the natural environment, including a great variety of bird species and other animals, as well as flowering plants. Plato (1817) praises Homer for integrating poetry with learning. Powell (Powell, 2007) considers Homer the "richest source of information about early Greece," while others attribute to him the origin of rhetoric (Knudsen, 2014) and political thought (Hammer, 2002).

Aspects of human psychology, such as passions, arrogance, pride, judgment, mental power, intelligence, and dreaming experiences take center stage in the epics. Medical matters, such as traumatic bodily injuries suffered by warriors in the battlefield, are prominently featured in the *Iliad*. These are addressed in several chapters of this book. The *Iliad* also describes an epidemic disease, attributed to god's wrath, that spread rapidly through the military camp with devastating consequences for the Greek army. A poison that enhances the lethality of arrows is mentioned in the *Odyssey*, along with an allusion to ethical constraints surrounding its use in war.

The poems contain information on the herbal pharmacology that existed in Homer's time. This involves medicinal herbs used for alleviating pain and suffering, or for softening grief and anxiety. They also narrate, in the context of mythic stories, the use of toxic drugs that alter mental functions and behavior, along with medicinal herbs capable of counteracting these effects (antidotes). The idea of searching for a specific antidote against a poisonous drug continued in the writings of Greco-Roman era botanists, including Theophrastus, Dioscorides, Pliny, and down to our times.

The Homeric poems are remarkable for their propensity to describe real-life events along with mythic/fantastic stories. Because these stories may be allegories hiding ancient wisdom, uncovering their true meaning could reveal precious knowledge that

Medicine in Homer. DOI: https://doi.org/10.1016/B978-0-443-32978-4.00001-8

existed in the remote past. While this has remained challenging since antiquity, recent advances in medical and natural sciences offer the opportunity to interpret the Homeric tales through the lens of current scientific evidence and thought. This is attempted here as detailed below.

Homeric epics and the power of allegory

The *Iliad* and *Odyssey* are the oldest poetic works of European literature. These are widely known as the Homeric epics, based on the belief that Homer composed them. They are monumental poetic creations, with the *Iliad* running 15,693 poetic lines and the *Odyssey* 12,109. The *Iliad* describes a major event that occurred during the Trojan War that began with an epidemic disease striking the Greek army and leading to the wrath of Achilles. The *Odyssey* deals with the wanderings of Odysseus on his venture to return home to Ithaka after the fall of Troy. The Homeric epics glorify personal sacrifice in the pursuit of honor and greatness. However, in addition to scenes of war and death, the epics also depict acts of civility and compassion, as well as many aspects of human life and the natural world.

Although criticized by some, Homer has been held in high esteem throughout the centuries, having inspired generations of writers and thinkers (Pfeiffer, 1968). During the Greco-Roman era, the epics were initially translated into Latin and, after the 16th century of the Common Era, into English, French, German, and other languages. In recent years, an increasing number of new translations of the *Iliad* and *Odyssey* have appeared, suggesting that, in spite of the immense changes that have occurred in modern societies, the Homeric epics still maintain considerable appeal.

The prevailing belief today is that the poems originated from the oral heroic tradition and passed on as such for centuries until written down during the 8th century BCE. However, there are still many uncertainties regarding these assumptions, with the centuries-old "Homeric Question" remaining essentially unresolved (Powell, 2007; West, 2011). While the epics are thought to describe the world of the protohistoric Iron Age of Greece, the *Iliad* refers to the Trojan War dated to the Late Bronze Age (12th century BCE). Moreover, the epics contain elements of the Mycenaean civilization that may not have been available to archaic Greeks (Nilsson, 1932). Powell (Powell, 2007) has accordingly acknowledged that "Homer's relationship to the Bronze Age remains a persistent problem."

As noted earlier, the epics contain knowledge about pharmacology and toxicology in conjunction with the natural world that existed in Homer's time. However, it is presently unclear to what extent this material derives from archaic Greece, or from the Late Bronze Mycenaean or even Minoan eras. Both civilizations had an advanced understanding of the natural world, as revealed in extant wall paintings and pottery that depict an amazing variety of plants and animals. Also, Nilsson (1932) has presented evidence for the Mycenaean origin of the ancient Greek religion as this became known in the Homeric epics and in Hesiod's *Theogony*. This view is supported by the subsequent decipherment of the Linear B script, used for writing in Mycenaean palaces, which depicts deities' names as they appear in the poems (Chadwick, 1990).

With the destruction of the Mycenaean centers in the 12th century BCE (Knodell, 2021), the heroic world vanished, to be followed by the Early Iron Age, known by some as the "Dark Age of Greece." During this period, population size decreased and living standards declined along with disruption of information transfer and trade (Dickinson, 2006; Snodgrass, 2000). Around the time of the Mycenaean collapse, destructive events also took place in other civilization centers in Asia Minor, such as the Hittite empire, and in the Middle East. There are many theories regarding the cause(s) of this nearly simultaneous widespread disintegration, including the possibility that communicable diseases may have played a role, given that epidemics respect no borders (Norrie, 2016). As such, the pestilence described in the Homeric epics may derive from the collective memory that existed in Homer's time of epidemics afflicting the Late Bronze or archaic world.

As the heroic age was succeeded by a dark period lasting up to four centuries, it is presently unclear how the Bronze Age knowledge survived in the Homeric epics. While the prevailing belief is that poems were preserved by the memory of bards, Nilsson (Nilsson, 1932) finds it difficult to believe that these survived for more than three generations (about one century). It should, however, be noted that unlettered people may be able to memorize and transmit information better than those who rely on writing as a means of communication (Ong, 2002; Plato, 1817; Sparrow et al., 2011). Parry (Parry, 1932, 1933) noted that the Homeric epics contain repetitive material or formulas that may have played an important role in the memorization and oral transmission of the poems.

A key feature of the Homeric epics is that knowledge acquired empirically and philosophical concepts are often presented in the context of mythic events. By using legends, rather than factual accounts, Homer is thought to convey historical truths and wisdom allegorically. These metaphors are likely intended to raise the interest of the audience, unlocking their imagination and intelligence. For example, turning men into pigs, as occurs in the Circe episode in the *Odyssey*, likely conveys the hidden message that behavior is what actually defines a human versus an animal (see below). Indeed, the art of using allegory constitutes a powerful means of expression in Homer, one which has had a lasting influence on thinkers, philosophers and playwrights throughout the centuries (Konstan et al., 2022; Russell, 2003; Tate, 1927). The messages hidden in Homeric allegory may have led to the subsequent development of cosmology and philosophy, as initially postulated by Theagenes of Rhegium (6th century BCE), an early interpreter of the Homeric tales. Hence, uncovering the meaning of the Homeric legends, as attempted here, may help us to better understand the early Greek civilization and its enduring influence.

Although the historic value of the Homeric poems has been debated, many scholars find that the epics contain essentially accurate information on a host of topics, including human anatomy and traumatic bodily injuries (covered extensively in this book), the natural environment (Brockliss, 2019), seamanship technology (Casson, 1968) and astronomical phenomena (Baikouzis & Magnasco, 2008). In support of the view that Homer's tales are based on history are startling archeological discoveries made over the past one and a half centuries, partly on the basis of the Homeric political map (Powell, 2007). Accordingly, the present work stems from the premise that the Homeric epics are the legacy of advanced prehistoric or protohistoric societies, which contain valuable empirical knowledge acquired by these cultures.

As such, in interpreting the episodes examined here, all relevant details contained in the epics will be accounted for and a faithful translation, by the present author, of the Homeric passages cited here will be employed.

Homer's pharmacology and toxicology

Both the *Iliad* and *Odyssey* describe medicinal herbs used for treating wounds, as well as drugs that relieve pain, anxiety, emotional suffering and grief. In addition to therapeutic medicines, the epics also record toxic pharmaceutical agents allegedly used to alter mental function and behavior, along with drugs capable of counteracting these effects. This knowledge is often presented metaphorically in the context of mythic stories, as mentioned earlier. Such Homeric allegory has stimulated human imagination since antiquity, with many naturalists and thinkers having attempted to identify the real substances behind the legendary drugs recorded in the epics (Ramesh Sankar, 2024; Stannard, 1962). Whereas various assumptions and exegeses have appeared in the literature over the centuries, the advent of modern scientific thought has led to exciting new opportunities to interpret the Homeric legends.

Circe and behavior-altering drugs

Among the fascinating stories presented allegorically in the *Odyssey* is the episode involving the *πολυφάρμακος* (*polypharmacist*) wise in drugs Circe, who uses medicinal herbs to enchant and control men. According to this story, Odysseus and his crew arrive in the island of Aeaea (*Αιαιη*), the home of the nymph Circe. After disembarking at the island's harbor, Odysseus' comrades led by Eurylochus walk through the woodlands and up the hills and find the great house of Circebuilt in a forest glade. When approaching the house, Odysseus' comrades are distressed by the sight of strangely behaved wild animals, including mountain wolves and lions which, according to the verse, Circe has tamed and pacified by giving them evil drugs.

Nevertheless, captivated by the melody of Circe's song, the men enter the stately home, except for Eurylochus who remains outside, suspecting deception. Circe, as was her custom, prepares for them a potion, blended with wine and mixed with evil drugs: (*Odyssey X, 230–237*): *ἀνέμισγε δὲ σιτῳ φάρμακα λύγρ', ινα πάγχυ λαθοιατο πατριδος αιης (but she mixed with the food pernicious drugs, to make them totally forgetful of their father land)*. (This and all translations by A. Plaitakis.) After drinking the potion and becoming utterly forgetful, Circe strikes the men with her magic wand, turning them into pigs. The narrative continues that the men, confined in pigsties, cried and rolled around in the muddy ground like pigs but retained their human minds.

Eurylochus returns to the harbor and breaks the bad news to Odysseus, who promptly leaves the ship behind and walks up the hill to rescue his men. Then, while crossing a glade area (*OdysseyX, 275*): *ιὼν ιερὰς ἀνὰ βήσσας (when going through the sacral glades)*, he encounters the god Hermes disguised as a handsome young man. Hermes, after warning Odysseus of the dangers of the place (*Odyssey X, 281–282*): *πῇ δὴ αὖτ', ὦ δύστηνε, δι' ἄκριας ἔρχεαι οιος, χώρου ἄϊδρις ἐών; (where are you going, unfortunate man, on these heights being ignorant of the land?)*, provides him with a wonderful medicine that will protect him from Circe's poisonous drugs. Specifically, Hermes tells Odysseus: (*Odyssey X, 290–292*): *τεύξει τοι κυκεῶ, βαλέει δ' ἐν φάρμακα σιτῳ. ἀλλ' οὐδ' ὥς θέλξαι σε δυνήσεται· οὐ γὰρ ἐάσει φάρμακον ἐσθλόν, ὅ τοι δώσω. (she will prepare a potion and put drugs into the food, but she will not be able to charm you, because the good medicine I shall give will not permit it).* The narrative continues: (*Odyssey X, 302–306*):

ὥς ἄρα φωνήσας πόρε φάρμακον ἀργεϊφόντης,
ἐκ γαίης ἐρύσας, καί μοι φύσιν αὐτοῦ ἔδειξε.
ῥίζῃ μὲν μέλαν ἔσκε, γάλακτι δὲ εἴκελον ἄνθος·
μῶλυ δέ μιν καλέουσι θεοί· χαλεπὸν δέ τ' ὀρύσσειν
ἀνδράσι γε θνητοῖσι, θεοὶ δέ τε πάντα δύνανται.
(*so, saying, Argeiphontes provided the medicine*
pulling it from the earth and showed its nature to me.
It had a black root and a milk-like flower
the gods call it moly, it is hard for mortal men
to dig it up; but the gods are capable for all things)

Once Odysseus enters the beautiful house of the nymph, Circe serves him the drug-containing magic potion prepared in a golden cup. However, she is astonished to see that Odysseus, having drunk the potion, remains unaffected. She promptly expresses her surprise (*Odyssey X, 326–329*):

θαῦμά μ' ἔχει ὡς οὔ τι πιὼν τάδε φάρμακ' ἐθέλχθης·
οὐδὲ γὰρ οὐδέ τις ἄλλος ἀνὴρ τάδε φάρμακ' ἀνέτλη,
ὅς κε πίῃ καὶ πρῶτον ἀμείψεται ἕρκος ὀδόντων.
σοὶ δέ τις ἐν στήθεσσιν ἀκήλητος νόος ἐστίν.

wonder holds me that drinking of these drugs you were not charmed
as no other man who drunk these drugs ever endured them
once they had passed the teeth barrier;
but in your breast, your mind remains unenchanted.

Shaken by the unexpected experience, Circe yields to Odysseus' demands to liberate his men and return them to their normal human state. As such, the episode emphasizes that Homer values mental ingenuity over physical power. To restore the men, Circe uses a different medicine (antidote) than that she originally employed to make the men forgetful and transform them into animals. Hence, the *πολυφάρμακος (polypharmacist)* Circe is highly knowledgeable of the pharmacological effects of behavior-altering drugs. In fact, she is capable of using these agents at her will to keep animals and humans under her possession and dominance. As noted above, this sensational story has fascinated people since antiquity. Current thoughts on the real nature of these magic medicines are detailed below.

The nature of Circe's drugs

As Circe's drugs cause amnesia followed by a delusional-hallucinatory state, it has been logically proposed that these drugs were anticholinergic herbs known to ancient Greeks (Lewin, 1924). These include *Datura stramonium* (Greek: *στρύχνος μανικός* or *maddening strychnos*; known as *jimson weed*), *Hyoscyamusniger* (Greek: *υοσκύαμος* from *ὗς+κύαμος "hog"+"bean"*; known as *henbane* or *stinking nightshade*) and *Atropa belladonna*. The term *Atropa* derives from Atropos, the eldest of the Three Fates who reigned over human destiny. Atropos, meaning the *inexorable* or *irreversible*, had the duty of cutting the thread of life. Interestingly, *Atropa belladonna* is known in English as *deadly nightshade*.

These poisonous herbs block the muscarinic acetylcholine receptor (Fig. 17.1), inducing a *central anticholinergic intoxication* (Jennings, 1935). In its advanced stage, the *anticholinergic syndrome* involves a plethora of neuro-psychological symptoms, including amnesia, delusions, hallucinations, blurred vision, unintelligible speech, and difficulty in walking. Inability to walk could have caused the men to crawl in the muddy soil, thus displaying a pig-like behavior, and their incoherent speech may have been perceived as a pig grunt. Hence, transformation of men into pigs while maintaining their human mind may in fact be an allegorical concept conveying the message that behavior is what defines a human being as compared to an animal.

The antidote "moly" identified as Galanthus nivalis *(snowdrop)*

It has been argued that flowers featured in poems are those known to the society from which the poetry originates (Brockliss, 2019). As such, the miraculous flower "moly"

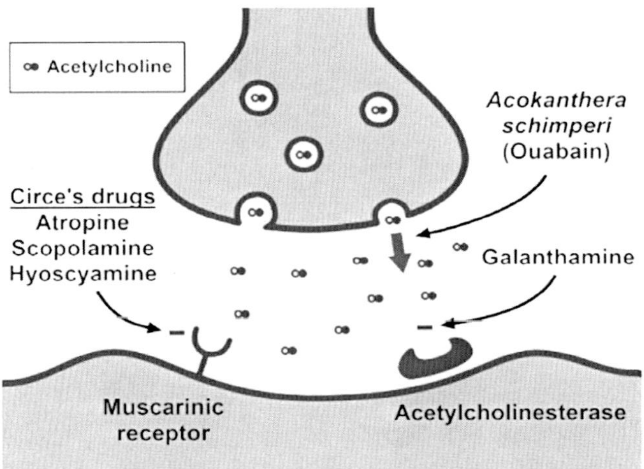

Figure 17.1 Schematic representation of a cholinergic synapse. Acetylcholine released from presynaptic vesicles acts on postsynaptic muscarinic receptor mediating synaptic transmission. Synaptic action is terminated by hydrolysis of acetylcholine via acetylcholinesterase. Circe's drugs are thought to induce central anticholinergic intoxication by blocking the acetylcholine muscarinic receptor. Inhibition of acetylcholinesterase by galanthamine, contained in the mythic plant "moly" (*Galanthus*), increases the levels of acetylcholine which in turn displace the toxic drugs from the muscarinic receptor. The arrow poison described in the text below is thought to be *Acokanthera schimperi*, a toxic plant containing cardiac glycosides, including ouabain. These enhance the presynaptic release of acetylcholine, resulting in bradycardia and even cardiac standstill.

was likely part of the flora that existed in Greece in Homer's time. There is evidence that this flora has not changed substantially since antiquity (Brockliss, 2019). While the identity of the antidote used by Odysseus to evade Circe's drugs has remained an enigma, Plaitakis and Duvoisin (Plaitakis & Duvoisin, 1983) proposed that the flower "moly" is likely to be *G. nivalis* (commonly known as snowdrop). This interpretation is based on the natural characteristics of *Galanthus*, which fit those of "moly," and the fact that the plant contains galanthamine, a naturally occurring acetylcholinesterase inhibitor that can reverse *central anticholinergic intoxication*. This thesis has been largely accepted by scholars and scientists alike (Ramesh Sankar, 2024; Lee, 1999; Pearson, 2001; Moss, 2020; Metz & Pavlov, 2021; Hensley et al., 2022) and has been cited by *The New York Times* (*Homer's Pharmacology, New York Times, Science Watch, Section C, page 3, June 7, 1983*). However, interpretation of Homer's point that the flower "is difficult for mortal men to dig it up" still remains an issue (Ramesh Sankar, 2024). Accordingly, this matter has been further addressed here (see below).

Regarding the natural characteristics of "moly," the Homeric text describes: *ῥιζῃ μὲν μέλαν ἔσκε, γάλακτι δὲ εικελον ἄνθος (has a black root and a milk-like flower).* The naturalists Theophrastus and Dioscorides (see below) added that "moly" has a "bulbous-like" root and that its leaves are on the ground. These details are consistent with the identification of "moly" as *Galanthus*, given that the plant has a dark-colored scallion root, from which the leaves emerge (Fig. 17.2). Also, the flower of *Galanthus* is pure white in color, like that of milk (Fig. 17.2). Of particular interest is that Carl Linnaeus, in 1753, coined the name *Galanthus* to denote the milk-like color of its blossom (Greek: gala (*γάλα*)=milk; anthos (*ἄνθος*)=flower). Whether Linnaeus knew of Homer's "moly" or it was a mere coincidence is unclear.

Regarding its pharmacologic profile, *Galanthus* harbors in all of its parts the alkaloid galanthamine, a tertiary amine capable of reversing anticholinergic intoxication by inhibiting acetylcholinesterase (Irwin & Smith, 1960; Mashkovsky, 1955). Acetylcholinesterase is an enzyme present in the synaptic cleft of cholinergic synapses that breaks down acetylcholine, an action that in turn terminates cholinergic transmission (Fig. 17.1). Inhibition of acetylcholinesterase by galanthamine increases synaptic acetylcholine levels, thereby overcoming the blockage of the muscarinic receptor induced by Circe's drugs (Plaitakis & Duvoisin, 1983). Because galanthamine is resistant to hydrolysis in body fluids, the drug is able to produce a long-lasting inhibition of acetylcholinesterase (Westra et al., 1986). Moreover, the ability of galanthamine to penetrate into the brain (Bickel et al., 1991) allows the drug to reverse anticholinergic intoxication (Baraka & Harik, 1977).

Cholinergic transmission plays a major role in central and peripheral nervous system functions. In the brain, cholinergic neurons that originate from the basal forebrain bundle project to the cerebral cortex (Deurveilher & Semba, 2011), where they are involved in learning, memory, and wakefulness (Hasselmo, 2006; Hut & Van der Zee, 2011). Loss of basal forebrain bundle cholinergic neurons occurs in Alzheimer's and other neurodegenerative diseases and is thought to contribute to the memory deficits of individuals affected by these disorders (Schliebs & Arendt, 2011). In these patients, enhancement of cholinergic transmission by galanthamine and other acetylcholinesterase inhibitors is found beneficial (Heinrich & Teoh, 2004).

Figure 17.2 *Galanthus nivalis* characteristics and comparison with *Leucojum* and *Scilla*. (A) *Galanthus nivalis* with its oblong-shaped white blossom (before it opens) standing on top of the leafless thin stalk. (B) The *Galanthus'* scallion root is dark brown in color approaching black in shade. (C) *L. vernum* (snowflake), the flower of which is similar, but larger than that of *Galanthus*. This is consistent with Dioscorides' description of "moly" (see text). (D) *Scilla bifolia*, the leaves of which are similar to *Galanthus*. See text regarding Theophrastus' description of "moly."
Source: Reproduced with permission from Plaitakis, A., & Duvoisin, R. C. (1983). Homer's moly identified as Galanthus nivalis L.: Physiologic antidote to stramonium poisoning. *Clinical Neuropharmacology, 6* (1), 1–5, https://doi.org/10.1097/00002826-198303000-00001.

Theophrastus and Dioscorides on the Homeric "moly"

As noted above, the identification of "moly" as *Galanthus* is supported by the writings of the ancient naturalists Theophrastus (4th century BCE) and Dioscorides (1st century CE). Their interest in "moly" suggests that Circe's story remained alive in the collective memory of the Greek people long after Homer's floruit. Specifically, Theophrastus in his *Historia Plantarum (9.15.7)* writes

τὸ δὲ μῶλυ περὶ Φενεὸν καὶ ἐν τῇ Κυλλήνῃ.
φασὶ δ' εἶναι καὶ ὅμοιον ᾧ ὁ Ὅμηρος εἴρηκε,
τὴν μὲν ῥίζαν ἔχον στρογγύλην προσεμφερῆ κρομύῳ
τὸ δὲ φύλλον ὅμοιον σκίλλῃ·

χρῆσθαι δὲ αὐτῷ πρός τε τὰ ἀλεξιφάρμακα καὶ τὰς μαγείας·
οὐ μὴν ὀρύττειν γ' εἶναι χαλεπόν, ὡς Ὅμηρός φησι
(the moly is around Pheneus and Kyllene,
They say to be similar to that Homer spoke about,
having a round root akin to an onion
and a leaf similar to that of Skilla.
it is used to against poisonous drugs and magic spells
but is not difficult to dig it up, as Homer says)

Kyllene is a forested mountain in the central Peloponnese and Pheneus is an adjacent valley crossed by rivers descending from the mountains. This natural environment is similar to the milieu of "moly" described in the *Odyssey* includes forest glades located at the heights (*ἄκριας*) of Circe's Island, where rivers flow from mountains down to the sea. The word *ἄκριας* (the plural of *ἄκρις* or *ἄκρος*, Latinized: *acris* or *acros*) gave rise to the modern term *acrophobia* meaning intense fear of heights. Also, his remark that the leaves of "moly" are similar to those of *Skilla* (Latinized: *Scilla*) supports the identification of "moly" as *Galanthus* (Fig. 17.2). Theophrastus is thought to have described different varieties of *Scilla*, that may include the *Scilla bifolia*, the leaves of which resemble those of *Galanthus* (Fig. 17.2). Lastly, Theophrastus' suggestion that "moly" is used against poisonous drugs agrees with Homer's testimony on the antidotal properties of the herb.

About four centuries after Theophrastus, Dioscorides (1st century CE), in his *De Materia Medica (3.47.1.1)* provides additional morphological details about "moly" and its relationship to *Leukoion (Leucojum)*:

Μῶλυ (οἱ δὲ λευκόϊον ἄγριον καλοῦσι) τὰ μὲν φύλλα
ἔχει ἀγρώστει ὅμοια, ἐπὶ γῆν ὄντα· ἄνθη λευκοΐοις
ὅμοια, γαλακτόχροα ἧσσονα δὲ πρὸς τὰ τοῦ ἴου·
καυλὸν δὲ λεπτὸν, πήχέων τεσσάρων· ἐπ' ἄκρου
δὲ ἔπεστιν, ὡς σκοροδοειδές· ἡ ῥίζα δὲ μικρά, βολβοειδής ·
Moly (them call it wild leukoion) has leaves similar
to those of grass, being on the earth, the flowers are like
those of leukoion, of milk color, but smaller than those of ion
a thin stalk of four cubits; on top stands (the flower)
resembling garlic, and the root is small, bulbous-like

Dioscorides' remark that "moly" is called *wild Leukoion* supports the identification of "moly" as *Galanthus*, as the two plants are very similar often being confused (Fig. 17.2). Indeed, molecular genetic analyses have confirmed the close phylogenetic relationship between them. There are even suggestions that *Galanthus* and *Leucojum* "could be lumped into a single genus" (Lledó et al., 2004). Dioscorides' observation that the flowers of "moly" are smaller than those of *Leukoion* is also in accord with this identification (Fig. 17.2). Lastly, the description that the root of "moly" is bulbous-like and its leaves are on the ground provides additional evidence that the magical plant is *Galanthus* as noted above.

In another passage in *De Materia Medica* (*3.123.1.1*), Dioscorides acknowledges that *Leukoion* was well known in his time (*λευκόϊον γνώριμον*). Indeed, the plant had been described earlier by Theophrastus (4th century BCE) who named it *λευκόιον*

(Leukoion) after the color of its flower (from leukos (*λευκός*)=white and ion (*ιόν*) =violet). Specifically, in his *Historia plantarum (7.13.9)*, Theophrastus specifies of *Leukoion*: *έπει τοις γε χρώμασι λευκά) (when it comes to the color, these are white)*. *Leukoion* is Latinized as *Leucoium* (Bauhini, 1623) from which the modern name *Leucojum* derives. In another passage in *Historia Plantarum* (6.8.1) Theophrastus observes

"Τῶν δ' ἀνθῶν το μέν πρῶτον ἐκφαίνεται τό λευκόϊον,
ὅπου μέν ό ἀὴρ μαλακώτερος εὐθύς τοῦ χειμῶνος,
ὅπου δὲ σκληρότερος ὕστερον, ἐνιαχοῦ τοῦ ἦρος"
(Of the flowers, the first to appear is the Leukoion,
where the air (climate) is milder, as soon as the winter starts,
but, where the air (climate) is harsher, in the spring in some places).

This seasonal flourishing fits that of *Leucojum vernum* (*Leucojum* of the spring) (Linnaeus, 1753), as well as the seasonal blooming of *Galanthus*, cheered as the *early flowering harbinger of spring* (Rønsted et al., 2013).

The natural habitat of Galanthus *is similar to that of the Homeric "moly"*

As noted above, Homer specifies that Odysseus obtained "moly" while crossing a forest glade (*βήσσας*) up in the heights of an island rich in oak trees, meadows, groves, springs, and rivers. The milieu of the Homeric "moly" fits the preferred ecological environment of *Galanthus*, which flourishes in an elevated terrain (300–800 m above sea level) in the moist, sheltered ground of forest glades, glens, meadows, and ravines (Figs. 17.3 and 17.4).

Recently we encountered *Galanthus* flowers during a casual walk toward the Vikos Gorge in the Epirus region of Greece (elevation about 800 m above sea level) (Fig. 17.3). In this mountainous wooded area, *Galanthus* grows wild. However, because of its small size, this inconspicuous flower can be easily missed if not searching carefully for it. This supports the interpretation of the Homeric verse

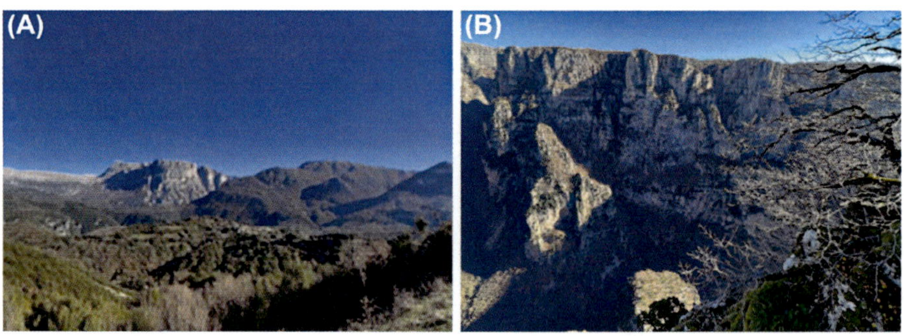

Figure 17.3 Mount *Tymphe* and Vikos Gorge. (A) The landscape around Mount *Tymphe* (northern Pindus Mountain range) in the Epirus region of Greece, where the *Galanthus* flowers (snowdrop) shown in Fig. 17.4 were encountered. (B) The adjacent Vikos Gorge.

Figure 17.4 *Galanthus* flowers in their natural habitat in Epirus, Greece. *Galanthus* flowers encountered during the first week of January 2023 near the Monodendri village in the Mount Tymphe of Epirus on a casual walk toward Vikos Gorge. (A) *Galanthus* flowers, growing among dry leaves and branches, are difficult to discern with the naked eye. (B) Close-up photo showing *Galanthus* flowers displaying their pure white bell-shaped blossoms. (C) Close-up photo showing *Galanthus* flowers with their oblong-shaped milk-white blossoms prior to their opening. (D) A *Galanthus* flower growing in a shady place next to a stone formation.

as actually meaning that "moly" is difficult to find, rather than to dig up (Plaitakis & Duvoisin, 1983). This interpretation also accords with Theophrastus' point (see above) that, contrary to Homer, "moly" is not difficult to dig up. His opinion carries weight, given that, in his *De Causis Plantaranum* (*Περι φυτικῶν αιτιῶν*), he systematically studied the role of the root in plants' physiology.

The chance finding of *Galanthus* in the Tymphe mountain of Epirus suggests that the plant is more ubiquitous in Greece than previously appreciated. This is not surprising, as the natural environment of Greece (a country covered 80% by mountains), along with its wet winters and moist springs, are suitable for *Galanthus* flourishing. Similar climatic conditions have been described by Theophrastus (4th century BCE) in his *Historia Plantarum* and by Hesiod (8th to 7th centuries BCE) in his *Works and Days* (Brockliss, 2019). In addition to the mainland, this ecological environment is encountered in the Greek islands, including the Heptanese, a cluster of seven islands in the Ionian Sea, and Sporades, a group of four main islands in the Aegean Sea. Elevated wooded areas also occur on the island of Samothrace (in the northern Aegean Sea) and in the islands of Ikaria and Samos (in the eastern Aegean Sea), where particular *Galanthus* species are endemic (see below).

Leucojum and Galanthus *through the centuries*

During the Greco-Roman era, Pliny (1st century CE) translated Theophrastus' *Leukoion* into Latin as *Viola alba*. In the late Renaissance, the Swiss botanist Caspar Bauhin (Bauhini, 1623) in his treatise *Pinax theatri botanici* placed Theophrastus' *Leukoion* under *Leucoium bulbosum vulgare* (Book II, Sect. 1, pages 55–56). Later, Linnaeus (Linnaeus, 1753) erected the *Leucojum* genus, placing *L. bulbosum vulgare* under *L. vernum*. He also included under this genus the *Leucojum autumnale* and *Leucojum aestivum*. Hence, the *Leucojumvernum* has descended to us through a historical line spanning over two millennia.

Linnaeus (Linnaeus, 1753) separated *Galanthus* from *Leucojum* on the basis of anatomical criteria. As noted above, the Linnean *Leucojum* genus included three species. However, Salisbury and Hooker (Salisbury & Hooker, 1805), moved *Leucojumautumnale* into the new genus of *Acis* (named after the myth of Acis and Galathea). Recent phylogenetic studies (Meerow et al., 2006; Larsen et al., 2010; Rønsted et al., 2013) support this classification by showing that *Galanthus*, *Leucojum*, and *Acis* are sister genera (Fig. 17.5A). All three belong to the Galantheae tribe of the Amaryllidaceae family.

At the time "moly" was identified (over 40 years ago) (Plaitakis & Duvoisin, 1983), *Galanthusnivalis* was regarded as the prevailing snowdrop species in Europe. Since then, however, an increasing number of *Galanthus* species have been recognized. At present, more than 20 *Galanthus* species are known, most of which concentrate in the Eastern Mediterranean, including the Greek mainland and the Aegean Islands. *Galanthus* originated from the Caucasus, from where over millions of years it dispersed to the neighboring regions (Davis, 2001, 1999; Rønsted et al., 2013).

The phylogenetic origin of *Galanthus* species that occur in the Greek landscape is shown in Fig. 17.5B. Of these, *Galanthus ikariae* was originally discovered in the Aegean Island of Ikaria (Baker, 1893) and is presently considered an endangered species (Davis, 2013). It is also endemic to other Aegean islands (Phitos et al., 1995; Dimopoulos et al., 2013). Phylogenetically, *G. ikariae* diverged from the *Galanthus elwesii* lineage as shown in Fig. 17.5B. *Galanthus trojanus* is found in western Asia Minor and is named after the city of Troy (Davis & Özhatay, 2001). The plant

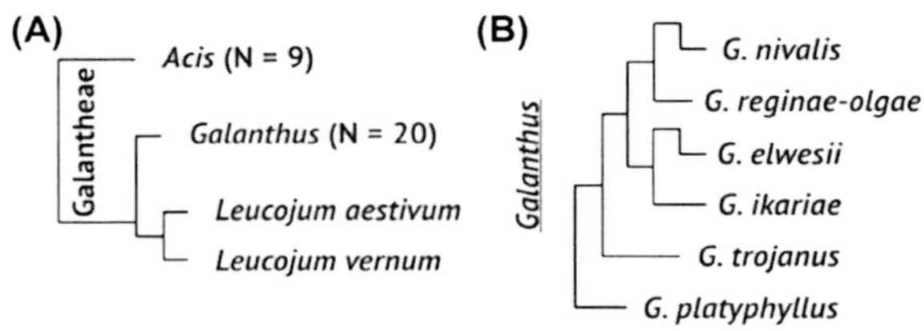

Figure 17.5 Phylogenetic relationships in the Galantheae tribe and the *Galanthus* genus. (A) Phylogenetic relationships in the Galantheae tribe belonging to the amaryllis family (Amaryllidaceae). As shown here, *Acis* separated from the common ancestor of *Leucojum* and *Galanthus* prior to their split (Rønsted et al., 2013). Hence *Galanthus* is more closely related to *Leucojum* than to *Acis*. (B) Phylogenetic relationships in the *Galanthus* genus. Selected species relevant to this work are shown. Based on Rønsted et al. (2013).

evolved rather early during the evolution of the *Galanthus* genus (Fig. 17.5B). *Galanthus reginae-olgae* was first described in Greece (in the year 1876) and named after the queen Olga of Greece (Beckett, 1993). Besides Greece, *G. reginae-olgae* also occurs in the Illyrian Sea and Sicily (Rønsted et al., 2013), having diverged from the lineage of *G. nivalis* (Fig. 17.5B). The latter is rather uncommon in Greece, although it shows a wide distribution elsewhere in Europe ranging from the Pyrenees to Southern, Central, and Eastern European countries (Davis, 2001, 1999). *Galanthus elwesii* is found in Turkey, Greece, and other countries of the Balkan Peninsula and in Odessa (Rønsted et al., 2013). Lastly, *Galanthus samothracicus* has been recently detected in the Aegean island of Samothrace (Tan et al., 2014), but its phylogenetic relationships have not as yet been determined.

Because *Galanthus* and *Leucojum* are quite similar and because both contain galanthamine, each of them is a candidate for the Homeric "moly." However, Dioscorides' observation (noted earlier) that the flower of "moly" is smaller than those of *Leukoion*, favors *Galanthus*. Regarding the identification of "moly" with a particular species of *Galanthus*, any that is endemic to the Aegean Islands (*G. ikariae* and *G. samothracicus*) is a strong candidate. Similarly, *G. reginae-olgae*, originally detected in Greece, is another strong candidate, being found in the Peloponnese where the main Mycenaean centers are located. Lastly, *Galanthus elwesii* and *Galanthus trojanus* found in western Asia Minor, need to be considered given that, according to some accounts, Homer was native to Ionia (Powell, 2007).

Alternative interpretations

John Goodyer in his translation of Dioscorides' *De Materia Medica* identified "moly" as *Allium nigrum*, apparently alluding to Dioscorides' description: *"ἐπ' ἄκρου δὲ ἔπεστιν, ὡς σκοροδοειδές" (on top it stands as garlic like)*. However, this refers to the blossom, which indeed stands on top of the thin stalk and which, before it opens, resembles garlic (Figs. 17.2 and 17.4). Nevertheless, Goodyer's identification was frequently cited in the literature over the past centuries. Modern scholars, however, do not accept this interpretation for the simple reason that Homer's "moly" does not resemble garlic (Lee, 1999). In recent years, additional interpretations have appeared in the literature (Ramesh Sankar, 2024; Stannard, 1962), but these also remain unsatisfactory (Ramesh Sankar, 2024).

Recently Molina-Venegas and Verano (2024) have reasoned that Homer's "moly" is unlikely to be *Galanthus* as, in their opinion, *Galanthus* is of Euro-Siberian affinity, being rare in the Mediterranean. Moreover, the authors suggest that the hot and dry climate of Greece is unsuitable for the growth of *Galanthus*. All of these assumptions, including the Euro-Siberian affinity of *Galanthus*, are clearly inaccurate. In fact, there is a convergence of scientific evidence that most *Galanthus* species concentrate in the Eastern Mediterranean, including the mainland and the islands of Greece where endemic *Galanthus* species occur. Moreover, the rainy wet winters and moist springs of Greece favor the flourishing of *Galanthus* well ahead of the dry hot summers, as illustrated in Fig. 17.4.

Furthermore, Molina-Venegas and Verano (2024) suggest *Pancratium maritimum L.* (known as sea daffodil) as an alternate candidate for "moly." *Pancratium maritimum L.*

indeed possesses anti-acetylcholinesterase activity, as several members of the Amaryllidaceae family do (Kaur et al., 2022). However, the plant grows preferentially in sand dunes at seaside. This habitat is inconsistent with the natural milieu of "moly," which, according to the Homeric text, was a forest glade away from the sea and up in the heights of Circe's Island as described earlier.

Helen's drug healing sorrow and passion

Telemachus, the son of Odysseus, in his efforts to learn the whereabouts of his father, visits king Menelaus in Sparta. During their conversation, the king recalls his suffering at sea and the anguish he felt when his brother, the great king Agamemnon, was murdered by his wife. Menelaus also expresses his sorrow at losing so many of his friends during the Trojan War; he frequently mourns his dead comrades and weeps often. Above all, however, he is painfully missing Odysseus, who had worked so hard for him. His words provoke the strong emotional reaction of Telemachus, whose tears roll down his cheeks. While everyone is mourning, Helen decides to act (*Odyssey IV, 219–230*):

ἔνθ' αὖτ' ἄλλ' ἐνόησ' Ἑλένη Διὸς ἐκγεγαυια
αὐτίκ' ἄρ' εις οινον βάλε φάρμακον, ἔνθεν ἔπινον,
νηπενθές τ' ἄχολόν τε, κακῶν ἐπίληθον ἀπάντων.
ὃς τὸ καταβρόξειεν, ἐπὴν κρητῆρι μιγείη,
οὔ κεν ἐφημέριός γε βάλοι κατὰ δάκρυ παρειῶν,
οὐδ' ει οι κατατεθναίη μήτηρ τε πατήρ τε,
οὐδ' ει οι προπάροιθεν ἀδελφεὸν ἢ φίλον υιὸν
χαλκῷ δηιόῳεν, ὁ δ' ὀφθαλμοισιν ὁρῷτο.
τοια Διὸς θυγάτηρ ἔχε φάρμακα μητιόεντα,
ἐσθλά, τά οι Πολύδαμνα πόρεν, Θῶνος παράκοιτις
Αιγυπτίη, τῇ πλειστα φέρει ζείδωρος ἄρουρα
φάρμακα, πολλὰ μὲν ἐσθλὰ μεμιγμένα πολλὰ δὲ λυγρά:

(*then Helen, daughter of Zeus, had another thought*
she immediately put into the wine they were drinking,
the drug nepenthes that removes anger, bringing forgetfulness of all bad things,
whoever swallows this drug down, when mixed in the bowl,
cannot during that day have a tear roll down his cheeks,
not even if his mother and father drop dead,
not even is a brother or dear son is killed before him
with deadly bronze, while seeing this with his own eyes.
such good and wise drugs had the daughter of Zeus
given to her by Polydamna, the wife of Thon
of Egypt, where the grain-giving earth holds drugs,
many are good when mixed, and many that are pernicious)

As narrated here, the drug *nepenthes* removes wrath and obliterates all painful memories. *Nepenthes* (*νηπενθές*) derives from the Greek words: *νη*=not+*πένθος*=sorrow, grief, or mourning (Merriam-Webster Dictionary). Theophrastus (9.15.1) also cites Homer saying that *nepenthes* eliminates rage: *ωστε λήθην ποιείν και απάθεια κακών (by inducing forgetfulness and insensibility to bad things)*. However, the naturalist does not attempt to identify *nepenthes*.

There are many speculations as to the nature of the drug that made the men calm, insensible, and forgetful of all bad happenings. Some believe that this could be the anticholinergic herb henbane (*Hyoscyamus niger*) (Hocking, 1947). While hyoscyamine can indeed cause amnesia, this is not selective for bad memories. Also, the drug induces behavioral changes, including hallucinations, delusions, and other neuropsychological manifestations, which Homer does not allude to. In fact, Menelaus and his guests, who drink the potion, are able to continue their lively conversation and other activities.

Over a century ago, *nepenthes* was identified as opium (Sprengel, 1817). This exegesis has been accepted by most modern pharmacologists (Lewin, 1924; Brownstein, 1993) and cited in *The New York Times* (*The Drug Helen Administered, New York Times, October 23, 1887*). This is indeed a reasonable interpretation, given that opium, extracted from the poppy plant *Papaver somniferum*, contains several psychoactive alkaloids (including morphine, codeine, and heroin), which interact with the μ–δ-opioid (Mu–Delta) receptor inducing analgesia, sedation, and euphoria. While euphoria is conventionally defined as a feeling of well-being and hopeful attitude, it can also lead to inappropriate psychological state and behavior. Although euphoria could account for the insensibility of those taking *nepenthes*, the indifference to family disasters, at least to the degree described in the *Odyssey*, is difficult to believe and, as such, it may represent poetic exaggeration.

It has been suggested (Kritikos & Papadaki, 1967) that those who drank opium in the palace of Menelaus in Sparta may not have experienced any of the toxic effects of opium because they used it habitually. Although it remains unclear on what evidence this assumption is based, recent clinical studies have shown that chronic opioid use does not substantially impair cognitive functions (Ersek et al., 2004), as apparently alluded to in the *Odyssey*.

As narrated in the *Odyssey*, *nepenthes* was provided to Helen by the Egyptian woman Polydamna. This seems likely, given that in ancient Egypt opium was used for medicinal purposes. In Egyptian mythology, opium is attributed to Thoth (the god of wisdom and knowledge), being given to Ra by Isis for the treatment of headache. While Sumerians are thought to be the first to cultivate opium poppies (around 3400 BCE), a method of preparing opium from poppy seeds is recorded in the Ebers Papyrus (around 1500 BCE). Hence, these considerations support the hypothesis that *nepenthes* was opium derived from Egypt, although poppies existed in the Greek landscape during the Bronze Age, having been depicted in extant Minoan artifacts.

The arrow poison

An episode described in the *Odyssey* refers to a poison sought by Odysseus to anoint his arrows. Homer calls the arrow poison *φάρμακονάνδροφόνον (mankiller drug)*. In this episode Athena, disguised as the Taphian king Mentes, visits Telemachus (the son of Odysseus) in Ithaka in order to encourage and inspire confidence in him.

Mentes recalls that he met Odysseus when he passed through the house of his father Anchialus in Taphos (thought to be located north of Ithaka) on his way back from Ephyra where he sought the arrow poison, as follows (*Odyssey I, 259–264*):

ἐξ Ἐφύρης ἀνιόντα παρ' Ἴλου Μερμερίδαο—
ᾤχετο γὰρ και κεισε θοῆς ἐπι νηὸς Ὀδυσσεὺς
φάρμακον ἀνδροφόνον διζήμενος, ὄφρα οι ειη
ιοὺς χρίεσθαι χαλκήρεας: ἀλλ' ὁ μὲν οὔ οι
δῶκεν, ἐπεί ῥα θεοὺς νεμεσίζετο αιὲν ἐόντας,

from Ephyra via the house of Ilus, son of Mermerus,
Odysseus was coming, having gone there with his swift ships
looking for a mankiller drug to smear the bronze head of arrows, but
Ilus did not give it to him, fearing the wrath of gods, who are forever

Hence, as king Ilus refused to give away the arrow poison by evoking the wrath of the gods, Homer recognizes ethical constraints on the use of such chemical warfare. Ilus is said to be living in the fabled city of Ephyra, which had been considered an imaginary place until recent excavations uncovered the city at the junction of the Kokytos and Acheron rivers, near the grove of Persephone, as pointed out by Homer (*Odyssey X 508–515*). Thucydides (1.46.4) describes that the city of Ephyra is situated in the Thesprotis region, in northwestern Greece, at the place indicated by Homer. While the excavated city of Ephyra is Hellenistic, archeologists also discovered Cyclopean walls and fragments of pottery that date back to the Mycenaean era (Fears & Stillwell, 1977).

The Homeric narrative raises important questions regarding the nature and the origin of the arrow poison. It suggests that the poison was not readily available in Greece. Indeed, Odysseus had to travel to northern Greece in search of it, with his request being refused by Ilus. Because the drug was hard to obtain, the poisonous herb was probably not native to Greece. Instead, it might have been obtained from abroad, as was the case for *nepenthes* (*νηπενθές*), brought to Sparta by Helen from Egypt. As most of the arrow poisons known today derive from tropical and subtropical regions, it is likely that the Homeric poison originated from faraway lands, such as Ethiopia, which was vaguely known to Greeks at the time of Homer. The Ethiopians (from *αιθοψ, Latinized Ethops =sun-burnt*) were first described by Homer as the furthest of men, who live at the world's end (*ἔσχατοι ἀνδρῶν*) (*Odyssey I, 24*) After Homer, ancient geographers placed Ethiopia south of Egypt.

Remarkably, Theophrastus in his *Historia Plantarum* (9.15.2) describes that in Ethiopia the arrows are smeared with a root that is *θανατηφόρος (deadly)* and that this is also practiced in Scythia (located north of Greece). Hort (Hort, 1916), who translated *Historia Plantarum* into English, speculated that the arrow poison derived from the *Acokanthera schimperi* tree (Apocynaceaefamily) that is native to Ethiopia, Eritrea, Somalia, and other eastern and central African countries. It contains cardiac glycosides, including ouabain, that can cause death by cardiac arrest.

Presently, several arrow poisons are known. These mainly derive from South America, Africa, and Asia, where they have been traditionally used by indigenous people during animal hunting and warfare. Of these, tubocurarine, originating from

South America, blocks cholinergic transmission at the neuromuscular junction. Historically, galanthamine was initially used for reversing muscle paralysis induced by tubocurarine during anesthesia (Mashkovsky, 1955). These observations were pivotal to the discovery of this drug and elucidating the mechanism of its action (inhibition of acetylcholinesterase).

The cardiac glycosides contained in *A. schimperi* inhibit the Na/K-ATPase sodium/potassium pump, leading to increased intracellular concentrations of Ca2+. In myocardial cells, excessive Ca2+ levels can cause increased myocardial excitability and eventually ventricular fibrillation resulting in death. In addition, in cholinergic neurons, increased Ca2+ levels can lead to enhanced release of acetylcholine from presynaptic endings (Fig. 17.1), which in turn can result in bradycardia and even cardiac standstill.

The epidemic in the *Iliad*

As noted in the Introduction, the *Iliad* describes an epidemic disease that spreads rapidly through the military camp during the last year of the Trojan War. The pestilence causes countless deaths and leads to a quarrel between Achilles and the king Agamemnon with dire consequences for the Achaeans. While the epidemic raged for nine days, Achilles, on the tenth day, calls the people in assembly, where he addresses the leader of the military expedition, Agamemnon, as follows (*Iliad I, 58–60*):

Ἀτρεΐδη νῦν ἄμμε παλιμπλαγχθέντας ὀΐω
ἂψ ἀπονοστήσειν, ει κεν θάνατόν γε φύγοιμεν,
ει δὴ ὁμοῦ πόλεμός τε δαμᾷ και λοιμὸς Ἀχαιούς:

Atreus's son, I think that now, having gone again astray,
we should return back home, if we would escape death
as both war and pestilence destroy Achaeans

Having said this, Achilles proposes to ask a seer (*μάντις*) or a priest (*ιερῆα*) or even a dream interpreter (*ὀνειροπόλος*), to find out the reason Apollo is angry with the Greeks. This task falls to the seer Kalchas, the best reader of birds as omens, who reveals that Apollo is in fact enraged against the king Agamemnon for dishonoring his priest Chryses. In revenge, Apollo, raging in his heart, came down from Olympus and inflicted a dreadful punishment upon the Achaeans. Specifically, the Homeric narrative (*Iliad I,10*) says: *νοῦσον ἀνὰ στρατὸν ὄρσεκακήν, ὀλέκοντοδὲλαοί, (raised a bad disease throughout the army, destroying the people).* Indeed, moving like a black night, Apollo approached the military camp and started shooting his toxic arrows (toxic derives from the Greek: *"toxon," the "bow"*) first at the animals and then at the men (*Iliad I, 48–52*):

ἕζετ' ἔπειτ' ἀπάνευθε νεῶν, μετὰ δ' ιὸν ἕηκε:
δεινὴ δὲ κλαγγὴ γένετ' ἀργυρέοιο βιοιο:,
οὐρῆας μὲν πρῶτον ἐπῴχετο και κύνας ἀργούς,
αὐτὰρ ἔπειτ' αὐτοισι βέλος ἐχεπευκὲς ἐφιεις
βάλλ': αιει δὲ πυραι νεκύων καίοντο θαμειαί.

then he sat away from the ships, and then he shot an arrow,
terrible was the twang of the silver bow,
at first, he assailed the mules and the swift dogs
and then he attacked the men themselves, shooting them with piercing arrow,
with numerous pyres of the dead burning continually.

Attribution of the pestilence to the wrath of Apollo reflects the prevalent belief in Homer's time that diseases represent divine punishment for human wrongs. Such concepts had dominated human life across cultures since prehistoric times (Vitiello et al., 2022). Notably, decipherment of the Hittite writings uncovered the plague prayers of the king Mursilis, who attributed the plague of the Hittite empire during the Bronze Age (1335 BCE) to god's punishment for a sin committed by his father (Karavites, 2008). These beliefs retained their hold in human societies for millennia after Homer.

In reality, however, the outbreak of an epidemic in the besieging Greek army is not an unexpected occurrence, given that war conditions provide a fertile ground for the infectious agent (contagion) to spread. Indeed, historical records from antiquity through modern times contain multiple occurrences of epidemics striking military camps during conflicts (Smallman-Raynor & Cliff, 2004; Norrie, 2016). These have been attributed to the crowded conditions and the poor hygiene in these camps.

While the nature of the infectious agent involved remains unknown, details provided by Homer may provide some clues about the origin of the pestilence. In this regard, the description that "numerous pyres of the dead were burning continually" throughout the camp during the first nine days of the epidemic suggests that the contagion was highly virulent and deadly, spreading swiftly among the people. Also, the Homeric notion that the disease first struck the domesticated animals and then the men suggest that the infectious agent caused an epizooty followed by an epidemic.

Because Homer does not describe any of the clinical manifestations of the epidemic that struck the Greek army, the likelihood of uncovering its cause is quite remote in the absence of biological evidence. Nonetheless, it may be useful to approach this daunting issue by taking into consideration the Homeric text in conjunction with the various infectious agents known to be operational during the Bronze Age and/or the Early Iron Age. In this regard, molecular genomic approaches using DNA isolated from victims buried in archeological sites have in recent years yielded important clues to the nature of past epidemics or pandemics and the evolutionary history of their pathogens (Spyrou et al., 2019). Some of these advances that are relevant to the Homeric epidemic are discussed below.

Yersinia pestis *or other bacterial pathogens*

Having said that, one thought that comes to mind is a form of plague caused by *Y. pestis*, a highly virulent contagion known to have infected Bronze Age Eurasian communities (Rasmussen et al., 2015; Spyrou et al., 2018) and thought to have contributed to the decline of prehistoric populations (Shennan et al., 2013). However, the Bronze Age *Y. pestis* differs from the modern pathogen strain responsible for the *bubonic plague* and, as such, its mode of transmission and pathogenicity remain unclear. In this regard, molecular genetic studies, using DNA from the teeth of human skeletons, have shown that the modern *Y. pestis* diverged from its Bronze Age ancestor by acquiring a new gene, the *Yersinia* murine toxin (*ymt*), and other evolutionary changes that permit the pathogen to be efficiently transmitted via an arthropod (flea) vector (Rasmussen et al., 2015).

Infected fleas transmit the disease by biting the skin of the victim, thus providing a portal for *Y. pestis* to enter the lymphatic system and cause regional lymphadenopathy. Because the fleas usually bite the skin of the lower extremities, the initial symptoms are swollen, painful lymph nodes in the groin, the hallmark of *bubonic plague* (*boubon* in Greek means the groin area). It is rapidly fatal, with most patients succumbing to the infection within four days. Another route of transmission is handling or eating meat of animals infected by *Y. pestis*. Indeed, recent studies have shown that the bacterium can infect a variety of domesticated animals, including goats (Christie et al., 1980), sheep (Dai et al., 2018), and dogs (Nichols et al., 2014). In turn, the epizootic disease can lead to human plague epidemics (Dai et al., 2018).

Notably, a *Y. pestis* strain capable of causing *bubonic plague* has been detected at an Armenian site that dates back to 951 BCE. This suggests that the disease might have originated at an earlier time in the Middle East, a view consistent with the hypotheses that the Hittite plague (c.1335 BCE) and the Philistine plague (c.1190 BCE) were caused by *Y. pestis* (Norrie, 2016; Vitiello et al., 2022); however, alternative interpretations have been proposed (Trevisanato, 2007). The *Y. pestis* strain discovered in Armenia is thought to have spread to Eurasia through migration of people and/or trade networks, eventually giving rise to strains that devastated Europe over the ensuing millennia. Indeed, genomic studies of DNA, obtained from the teeth of victims, have confirmed that *Y. pestis* was the cause of the Justinian Plague (7th century CE), the Black Death (14th century CE), and the Great Plague (17th century CE) (Rasmussen et al., 2015; Norrie, 2016).

Remarkably, *Y. pestis* genomes were recently recovered from human skeletons of Bronze Age Crete (4th to 2nd millennia BCE) (Neumann et al., 2022). However, the Cretan strain may not have adapted to efficient transmission via the flea vector (Neumann et al., 2022). It belongs to the LNBA lineage found throughout Eurasia and is considered of steppe-related ancestry (Neumann et al., 2022). Because this strain probably could not cause *bubonic plague*, it may have caused airborne *pneumonic plague* (Neumann et al., 2022), a form of the disease that is transmitted from person to person via respiratory droplets and that is rapidly fatal, with 90% of affected individuals dying within 1–2 days. However, the efficiency of transmission and the virulence of the Cretan *Y. pestis* strain remain unclear. Recently *pneumonic plague* was documented in mountain lions, with humans getting infected by handling their corpses (Wong et al., 2009).

In light of the above, the Homeric pestilence could have been a form of *plague* caused by *Y. pestis* that originated from the Middle East, given Troy's location in Asia Minor. Also, a Eurasian steppe origin is possible as the ancient city of Troy lies at the crossroads between Europe and Asia. The Homeric account that domestic animals were struck first raises the possibility of a zoonotic reservoir disease brought to the Greek military camp by infected animals. Whether a particular virulent strain of *Y. pestis* existed in Homer's time capable of causing a rapidly spreading epizooty followed by a highly lethal epidemic is presently unknown.

An alternative possibility is that the pestilence was typhus, which often affects people in military camps. Typhus represents a group of diseases caused by bacteria. It is of interest in this regard that *Salmonella enterica* genomes have been isolated from the Minoan Crete burial site along with the *Y. pestis* genomes described above (Neumann et al., 2022). The lineage of the *Salmonella enterica* isolated from Minoan Crete includes the Paratyphi C, Typhisuis, and Choleraesuis serovars. However, these ancient bacteria lacked a series of genes acquired by modern strains that convey enhanced toxicity along with the ability to evade the immune system (Neumann et al., 2022). As such, the virulence and transmission of these pathogens remains unclear.

Another possibility is tularemia, an infectious disease caused by the bacterium *Francisella tularensis*, which is thought to be responsible for plague epidemics during the Bronze Age (Trevisanato, 2007). The disease is transmitted to humans from infected animals via arthropod vectors, such as ticks and deer flies, or by direct contact with such animals. However, it does not spread directly from human to human. Tularemia has been suggested for the Hittite epidemic mentioned above as well as other epidemics that occurred in the Middle East in later times (Trevisanato, 2007). Because tularemia is a long-lasting disease, *F. tularensis* is not likely to have caused the short-lived Homeric pestilence.

Viral pathogens

Of the various viral agents implicated in human epidemics or pandemics, the *variola* virus is of historical interest as it produces skin lesions that can be depicted in works of art unearthed by archeologists. Also, smallpox lesions detected in the skin of mummies can provide material for disease confirmation. Indeed, the *variola* virus was recently recovered from the mummy of a child who lived between 1643 and 1665 CE in Europe. However, smallpox might have existed in Egypt almost two millennia earlier. Thus skin lesions compatible with smallpox have been observed in the face, neck, and upper extremities of the pharaoh Ramses V, who died in 1157 BCE (Simmons et al., 2015; Spinney, 2020). Since Egypt had established commercial and other links with the Helladic world, smallpox is a candidate for the Homeric epidemic. However, the human smallpox, as known today, does not affect animals in nature.

An alternative possibility is a rapidly spreading type of influenza associated with high human mortality. Such an exceptionally deadly influenza pandemic, commonly known as "Spanish flu," occurred during the years of 1918–1920 around the world resulting in millions of deaths. It was caused by an H1N1 subtype of influenza (Tumpey et al., 2005) which is not known to affect animals. On the other hand, rapidly spreading canine and equine influenza epizootics have been well documented (Crawford et al., 2006; Slater et al., 2014). The predominant causal virus (H3N8) of these epizootics is not ordinarily pathogenic to humans, although rare fatal cases have been reported (WION, 2023).

Recently Tsoucalas et al. (2014) published a comprehensive analysis of the Homeric pestilence. After considering various bacterial and viral agents, they concluded that alphaviruses were likely the "perfect candidates." In deriving this conclusion, the authors reason that the Greek military camp was located on the shores of the Scamander River, near a swampy area. This could have provided a breeding ground for mosquitos (the virus carriers) to multiply. Moreover, the authors describe that the alphaviruses can affect domestic animals. However, as noted by Tsoucalas et al. (2014), the alphaviruses are ordinarily found in the Western Hemisphere, with some strains having been isolated in Asia and even in Europe.

Obviously, the nature of the Homeric epidemic cannot be determined in the absence of biological evidence, such as that obtained for other pandemics that occurred millennia ago. One possibility to be considered, however, is that the epidemic described in the *Iliad* may have been caused by an infectious agent that existed in Homer's time but is no longer with us. As bacteria and viruses are subject to frequent evolutionary genetic modifications that can alter their virulence or can create a new pathogenic strain out of an innocuous one, any suggestion about the nature of the Homer epidemic that occurred three millennia ago remains highly speculative. However, given that this is a rapidly developing scientific field, the possibility of uncovering the true origin of the Homeric epidemic in the near future remains tenable.

The Homeric narrative relates that, after the Greeks returned the daughter of Apollo's priest to her father, the epidemic subsided. However, a rational explanation stands that the epidemic disappeared after it had run its course, probably due to the high death rate recorded in the *Iliad*, which may have diminished the pool of people susceptible to contagion. Also, the practice of burning the bodies of the diseased may have helped to stop the pestilence. As such, the epidemic described in the *Iliad* needs not depend on divine intervention, but rather on natural factors that were operational during the Trojan War.

Perspectives

In addition to describing the inhabited world and the natural environment, Homer narrates, often in the context of mythic stories, a multitude of human experiences, including medical and pharmacological matters. Exploring such clues, as we have done here, offers the opportunity to achieve insight into ancient knowledge and sophistication. However, because the Homeric poems describe both real-life events and legends, and because the boundaries between fiction and reality are often blurred, uncovering the historical truths and wisdom hidden in the Homeric tales remains a challenging and fascinating endeavor that needs further exploration.

Acknowledgments

I am indebted to Ian Knight for his invaluable help in preparing this manuscript, including the careful editing of the text and figure work, as well as for his valuable suggestions. I thank Iris Plaitakis for reviewing the manuscript and for her useful comments. I am also grateful to my wife Oana-Cristina for drawing my attention to the tiny and delicate snowdrop flowers scattered on the ground among dried leaves near the Vikos Gorge of Epirus Greece (depicted in

Fig. 17.4 of this chapter). I am also thankful to Maria Tourna, Head Librarian of the Blegen Library of the American School of Classical Studies of Athens, and to the Cataloging Librarian Eleni Sourligka, also at the Blegen, for their assistance in using the Library's resources.

References

Baikouzis, C., & Magnasco, M. O. (2008). Is an eclipse described in the Odyssey? *Proceedings of the National Academy of Sciences of the United States of America, 105*(26), 8823–8828. https://doi.org/10.1073/pnas.0803317105, http://www.pnas.org/content/105/26/8823.full.pdf.

Baker, J. G. (1893). A synopsis of the genera and species of Museae. *Annals of Botany, os-7*(2), 189–222. https://doi.org/10.1093/aob/os-7.2.189.

Baraka, A., & Harik, S. (1977). Reversal of central anticholinergic syndrome by galanthamine. *JAMA: The Journal of the American Medical Association, 238*(21), 2293–2294. https://doi.org/10.1001/jama.1977.03280220061024.

Bauhini, C. (1623). *Πιναξ (Pinax) Theatri Botanici Sive Index in Theophrasti Dioscorides Plinii Et Botanicorun*. Impensis Joannis Regis.

Beckett, K. (1993). Galanthus. *Encyclopaedia of Alpines, 1*, 510–520.

Bickel, U., Thomsen, T., Fischer, J. P., Weber, W., & Kewitz, H. (1991). Galanthamine: Pharmacokinetics, tissue distribution and cholinesterase inhibition in brain of mice. *Neuropharmacology, 30*(5), 447–454. https://doi.org/10.1016/0028-3908(91)90005-v.

Brockliss, W. (2019). *Homeric imagery and the natural environment*. Harvard University Press.

Brownstein, M.J. (1993). A brief history of opiates, opioid peptides, and opioid receptors. 90 (12), 5391–5393. https://doi.org/10.1073/pnas.90.12.5391.

Chadwick, J. (1990). *The decipherment of linear B*. Cambridge University Press, 1–174.

Christie, A. B., Chen, T. H., & Elberg, S. S. (1980). Plague in camels and goats: Their role in human epidemics. *Journal of Infectious Diseases, 141*(6), 724–726. https://doi.org/10.1093/infdis/141.6.724.

Crawford, C., Dubovi, E.P. J., Donis, R.O., Castleman, W.L., Gibbs Hill, R.C., Katz, J.M., Ferro, P., & Anderson, T.C. (2006). Proceedings of the North American Veterinary Conference Canine Influenza Virus Infection.

Dai, R., Wei, B., Xiong, H., Yang, X., Peng, Y., He, J., Jin, J., Wang, Y., Zha, X., Zhang, Z., Liang, Y., Zhang, Q., Xu, J., Wang, Z., Li, W., & Raoult, D. (2018). Human plague associated with Tibetan sheep originates in marmots. *PLoS Neglected Tropical Diseases, 12*(8), e0006635. https://doi.org/10.1371/journal.pntd.0006635.

Davis, A.P. (2013). Galanthus ikariae. The IUCN red list of threatened species.

Davis, A. P., & Özhatay, N. (2001). Galanthus trojanus: a new species of Galanthus (Amaryllidaceae) from north-western Turkey. *Botanical Journal of the Linnean Society, 137*(4), 409–412. https://doi.org/10.1111/j.1095-8339.2001.tb02335.x.

Davis, A. P. (1999). *The genus Galanthus. A botanical magazine monograph*. Timber Press.

Deurveilher, S., & Semba, K. (2011). Basal forebrain regulation of cortical activity and sleep-wake states: Roles of cholinergic and non-cholinergic neurons. *Sleep and Biological Rhythms, 9*(1), 65–70. https://doi.org/10.1111/j.1479-8425.2010.00465.x.

Dickinson, O. (2006). *The Aegean from Bronze Age to Iron Age: Continuity and change between the twelfth and eighth centuries BC*. Routledge Taylor & Francis Group, 1–298. http://www.taylorandfrancis.com/books/details/9780203968369, https://doi.org/10.4324/9780203968369.

Dimopoulos, P., Raus, T., Bergmeier, E., Constantinidis, T., Iatrou, Kokkini, G., Strid, A., & Tzanoudakis, D. (2013). *Vascular plants of Greece. An annotated checklist*. Botanic Gardens and Botanical Museum Berlin-Dahlem and Hellenic Botanical Society, 1–372.

Ersek, M., Cherrier, M. M., Overman, S. S., & Irving, G. A. (2004). The cognitive effects of opioids. *W.B. Saunders, United States Pain Management Nursing, 5*(2), 75–93. https://doi.org/10.1016/j.pmn.2003.11.002, http://www.elsevier.com/inca/publications/store/6/2/3/1/0/7/index.htt.

Fears, J. R., & Stillwell, R. (1977). The Princeton encyclopedia of classical sites. *The American Historical Review, 82*(3), 611. https://doi.org/10.2307/1850902.

Hammer, D. (2002). *The Iliad as politics. The performance of political thought*. University of Oklahoma Press, 1–308.

Hasselmo, M. E. (2006). The role of acetylcholine in learning and memory. *Current Opinion in Neurobiology, 16*(6), 710–715. https://doi.org/10.1016/j.conb.2006.09.002.

Heinrich, M., & Teoh, H. L. (2004). Galanthamine from snowdrop - The development of a modern drug against Alzheimer's disease from local Caucasian knowledge. *Journal of Ethnopharmacology, 92*(2–3), 147–162. https://doi.org/10.1016/j.jep.2004.02.012.

Hensley, K., Danekas, A., Farrell, W., Garcia, T., Mehboob, W., & White, M. (2022). At the intersection of sulfur redox chemistry, cellular signal transduction and proteostasis: A useful perspective from which to understand and treat neurodegeneration. *Free Radical Biology and Medicine, 178*, 161–173. https://doi.org/10.1016/j.freeradbiomed.2021.11.039, http://www.elsevier.com/locate/freeradbiomed.

Hocking, G.M. (1947). Henbane: Healing herb of hercules and of Apollo.

Hort, A. (1916). *Translation of Theophrastus's enquiry into plants*. Loeb Classical Library.

Hut, R. A., & Van der Zee, E. A. (2011). The cholinergic system, circadian rhythmicity, and time memory. *Behavioural Brain Research, 221*(2), 466–480. https://doi.org/10.1016/j.bbr.2010.11.039.

Irwin, R. L., & Smith, H. J. (1960). Cholinesterase inhibition by galanthamine and lycoramine. *Biochemical Pharmacology, 3*(2), 147–148. https://doi.org/10.1016/0006-2952(60)90030-7.

Jennings, R. E. (1935). Stramonium poisoning. A review of the literature and report of two cases. *The Journal of Pediatrics, 6*(5), 657–664. https://doi.org/10.1016/S0022-3476(35)80060-7.

Karavites, P. (2008). *Homer and the Bronze Age. The reflection of humanistic Idealsin diplomatic practices*. Gorgias Press LLC.

Kaur, H., Chahal, S., Jha, P., Lekhak, M. M., Shekhawat, M. S., Naidoo, D., Arencibia, A. D., Ochatt, S. J., & Kumar, V. (2022). Harnessing plant biotechnology-based strategies for in vitro galanthamine (GAL) biosynthesis: A potent drug against Alzheimer's disease. *Plant Cell, Tissue and Organ Culture, 149*(1–2), 81–103. https://doi.org/10.1007/s11240-022-02229-0, http://www.springer.com/journal/11240.

Knodell, A. R. (2021). *Societies in transition in Early Greece: An archaeological history*. University of California Press, 1–363. https://www.ucpress.edu/book/9780520380530/societies-in-transition-in-early-greece, 10.1525/luminos.101.

Knudsen, R. A. (2014). *Homeric speech and the origins of rhetoric*. Johns Hopkins University Press, 1–230. http://muse.jhu.edu/books/9781421412276.

Konstan, D. (2022). Homeric Problems Heraclitus the Allegorist. In R. S. Smith, & S. M. Trzaskoma (Eds.). *The Oxford handbook of Greek and Roman mythography*Oxford University Press. https://doi.org/10.1093/oxfordhb/9780190648312.001.0001.

Kritikos, P. G., & Papadaki, S. P. (1967). The history of the poppy and of opium and their expansion in the antiquity in the eastern Mediterranean area. *UNODC-Bulletin on Narcotics, XIX*.

Larsen, M. M., Adsersen, A., Davis, A. P., Lledó, M. D., Jäger, A. K., & Rønsted, N. (2010). Using a phylogenetic approach to selection of target plants in drug discovery of acetylcholinesterase inhibiting alkaloids in Amaryllidaceae tribe Galantheae. *Biochemical Systematics and Ecology, 38*(5), 1026–1034. https://doi.org/10.1016/j.bse.2010.10.005.

Lee, M. R. (1999). The snowdrop (Galanthus nivalis) from Odysseus to Alzheimer's. *Proceedings of the Royal College of Physicians of Edinburgh EH, 11*.

Lewin, L. (1924). *Phantastica, narcotic and stimulant drugs*. Dutton.

Linnaeus, C. (1753). Species Plantarum: Exhibentes plantas rite cognitas, ad genera relatas, cum differentiis specificis, nominibus trivialibus, synonymis selectis, locis natalibus, secundum systema sexuale digestas. *Impensis. Laurentii Salvii*.

Lledó, M.ª D., Davis, A. P., Crespo, M. B., Chase, M. W., & Fay, M. F. (2004). Phylogenetic analysis of Leucojum and Galanthus (Amaryllidaceae) based on plastid matK and nuclear ribosomal spacer (ITS) DNA sequences and morphology. *Plant Systematics and Evolution, 246*(3-4). https://doi.org/10.1007/s00606-004-0152-0.

Mashkovsky, M. D. (1955). Effects of galanthamine on the acetylcholine sensitivity of skeletal musculature. *Farmakologiia i Toksikologia (Moskow), 18*.

Meerow, A. W., Francisco-Ortega, J., Kuhn, D. N., & Schnell, R. J. (2006). Phylogenetic relationships and biogeography within the Eurasian clade of Amaryllidaceae based on plastid ndhF and nrDNA ITS sequences: Lineage sorting in a reticulate area? *Systematic Botany, 31*(1), 42–60. https://doi.org/10.1600/036364406775971787, http://docserver.ingentaconnect.com/deliver/connect/aspt/03636445/v31n1/s6.pdf?expires=1141201730&id=26737007&titleid=10212&accname=Elsevier&checksum=3D2DEDDEC102C33C90-7CCCDD61F54FE2.

Metz, C. N., & Pavlov, V. A. (2021). Treating disorders across the lifespan by modulating cholinergic signaling with galantamine. *Journal of Neurochemistry, 158*(6), 1359–1380. https://doi.org/10.1111/jnc.15243, http://onlinelibrary.wiley.com/journal/10.1111/(ISSN)1471-4159.

Molina-Venegas, R., & Verano, R. (2024). The quest for Homer's moly: Exploring the potential of an early ethnobotanical complex. *Journal of Ethnobiology and Ethnomedicine, 20*(1). https://doi.org/10.1186/s13002-024-00650-7, https://ethnobiomed.biomedcentral.com.

Moss, D. E. (2020). Improving anti-neurodegenerative benefits of acetylcholinesterase inhibitors in Alzheimer's disease: Are irreversible inhibitors the future? *International Journal of Molecular Sciences, 21*(10), 3438. https://doi.org/10.3390/ijms21103438.

Neumann, G. U., Skourtanioti, E., Burri, M., Nelson, E. A., Michel, M., Hiss, A. N., McGeorge, P. J. P., Betancourt, P. P., Spyrou, M. A., Krause, J., & Stockhammer, P. W. (2022). Ancient Yersinia pestis and Salmonella enterica genomes from Bronze Age Crete. *Current Biology, 32*(16), 3641. https://doi.org/10.1016/j.cub.2022.06.094, http://www.elsevier.com/journals/current-biology/0960-9822.

Nichols, M. C., Ettestad, P. J., VinHatton, E. S., Melman, S. D., Onischuk, L., Pierce, E. A., & Aragon, A. S. (2014). Yersinia pestis infection in dogs: 62 cases (2003-2011). *Journal of the American Veterinary Medical Association, 244*(10), 1176–1180. https://doi.org/10.2460/javma.244.10.1176, http://avmajournals.avma.org/doi/pdf/10.2460/javma.244.10.1176.

Nilsson, M. P. (1932). *The Mycenean origin of Greek mythology*. University of California Press, 1–258.

Norrie, P. (2016). *How disease affected the end of the Bronze Age*. Springer Science and Business Media LLC, 61–101. 10.1007/978-3-319-28937-3_5.

Ong, W. J. (2002). *Orality and literacy*. Routledge, 1–264.

Parry, M. (1932). Studies in the epic technique of oral verse-making: II. The Homeric language as the language of an oral poetry. *Harvard Studies in Classical Philology, 43*, 1. https://doi.org/10.2307/310666.

Parry, M. (1933). Whole formulaic verses in Greek and Southslavic Heroic Song. *Transactions and Proceedings of the American Philological Association, 64*, 179. https://doi.org/10.2307/283165.

Pearson, V. E. (2001). Galantamine: A new Alzheimer drug with a past life. *Annals of Pharmacotherapy, 35*(11), 1406–1413. https://doi.org/10.1345/aph.1A092, http://aop.sagepub.com/content/by/year.

Pfeiffer, R. (1968). *History of classical scholarship: From the beginnings to the end of the Hellenistic age*. Oxford University Press, 144.

Phitos, D., Strid, A., Snogerup, & Greuter, W. (1995). The red data book of rare and threatened plants of Greece. World Wide Fund for Nature.

Plaitakis, A., & Duvoisin, R. C. (1983). Homer's moly identified as Galanthus nivalis L.: Physiologic antidote to stramonium poisoning. *Clinical Neuropharmacology, 6*(1), 1–5. https://doi.org/10.1097/00002826-198303000-00001.

Plato. (1817). *Phaedrus*. Taxila Publications.

Powell, B. B. (2007). *Homer*. Blackwell Publishing Ltd, 1–240.

Ramesh Sankar, N. (2024). Potatosoap": Moly in James Joyce's Ulysses. *ANQ - Quarterly Journal of Short Articles Notes and Reviews, 37*(1), 49–54. https://doi.org/10.1080/0895769X.2022.2059752.

Rasmussen, S., Allentoft, M. E., Nielsen, K., Orlando, L., Sikora, M., Sjögren, K. G., Pedersen, A. G., Schubert, M., Van Dam, A., Kapel, C. M. O., Nielsen, H. B., Brunak, S., Avetisyan, P., Epimakhov, A., Khalyapin, M. V., Gnuni, A., Kriiska, A., Lasak, I., Metspalu, M., ... Willerslev, E. (2015). Early divergent strains of Yersinia pestis in Eurasia 5,000 years ago. *Cell, 163*(3), 571–582. https://doi.org/10.1016/j.cell.2015.10.009, http://www.cell.com.

Russell, D. (2003). The rhetoric of the Homeric problems. In G. R. Boys-Stones Metaphor (Ed.). *Allegory, and the classical tradition: Ancient thought and modern*. Oxford University Press.

Rønsted, N., Zubov, D., Bruun-Lund, S., & Davis, A. P. (2013). Snowdrops falling slowly into place: An improved phylogeny for Galanthus (Amaryllidaceae). *Molecular Phylogenetics and Evolution, 69*(1), 205–217. https://doi.org/10.1016/j.ympev.2013.05.019.

Salisbury, R.A., & Hooker, W. (1805). The Paradisus Londonensis: Or coloured figures of plants cultivated in the vicinity of the metropolis.

Schliebs, R., & Arendt, T. (2011). The cholinergic system in aging and neuronal degeneration. *Behavioural Brain Research, 221*(2), 555–563. https://doi.org/10.1016/j.bbr.2010.11.058.

Shennan, S., Downey, S. S., Timpson, A., Edinborough, K., Colledge, S., Kerig, T., Manning, K., & Thomas, M. G. (2013). Regional population collapse followed initial agriculture booms in mid-Holocene Europe. *Nature Communications, 4*. https://doi.org/10.1038/ncomms3486.

Simmons, B. J., Falto-Aizpurua, L. A., Griffith, R. D., & Nouri, K. (2015). Smallpox: 12000 years from plagues to eradication: A dermatologic ailment shaping the face of society. *JAMA Dermatology, 151*(5), 482. https://doi.org/10.1001/jamadermatol.2014.4812, http://archderm.jamanetwork.com/article.aspx?articleid=2119342.

Slater, J., Borchers, K., Chambers, T., Cullinane, A., Duggan, V., Elton, D., Legrand, L., Paillot, R., & Fortier, G. (2014). Report of the International Equine Influenza Roundtable Expert Meeting at Le Touquet, Normandy, February 2013. *Equine Veterinary Journal, 46*(6), 645–650. https://doi.org/10.1111/evj.12302.

Smallman-Raynor, M. R., & Cliff, A. D. (2004). Impact of infectious diseases on war. *Infectious Disease Clinics of North America, 18*(2), 341–368. https://doi.org/10.1016/j.idc.2004.01.009.

Snodgrass, A. M. (2000). *The Dark Age of Greece*. New York: Routledge, 1–456.

Sparrow, B., Liu, J., & Wegner, D. M. (2011). Google effects on memory: Cognitive consequences of having information at our fingertips. *Science (New York, N.Y.), 333*(6043), 776–778. https://doi.org/10.1126/science.1207745, http://www.sciencemag.org/content/333/6043/776.full.pdf.

Spinney, L. (2020). Smallpox and other viruses plagued humans much earlier than suspected. *Nature, 584*(7819), 30–32. https://doi.org/10.1038/d41586-020-02083-0.

Sprengel, K. (1817). Geschichte der Botanik. *Kurt Polycarp Joachim Sprengel, 1*.

Spyrou, M. A., Bos, K. I., Herbig, A., & Krause, J. (2019). Ancient pathogen genomics as an emerging tool for infectious disease research. *Nature Reviews. Genetics, 20*(6), 323–340. https://doi.org/10.1038/s41576-019-0119-1, http://www.nature.com/reviews/genetics.

Spyrou, M. A., Tukhbatova, R. I., Wang, C. C., Valtueña, A. A., Lankapalli, A. K., Kondrashin, V. V., Tsybin, V. A., Khokhlov, A., Kühnert, D., Herbig, A., Bos, K. I., & Krause, J. (2018). Analysis of 3800-year-old Yersinia pestis genomes suggests Bronze Age origin for bubonic plague. *Nature Communications, 9*(1). https://doi.org/10.1038/s41467-018-04550-9, http://www.nature.com/ncomms/index.html.

Stannard, J. (1962). The plant called Moly. *Osiris, 14*, 254–307. https://doi.org/10.1086/368629.

Tan, K., Biel, B., & Siljak-Yakovlev, S. (2014). Galanthus samothracicus (Amaryllidaceae) from the island of Samothraki, northeastern Greece. *Phytologia Balcanica, 20*(1), 65–70.

Tate, J. (1927). The beginnings of Greek Allegory. *The Classical Review, 41*(6), 214–215. https://doi.org/10.1017/s0009840x0003883x.

Trevisanato, S. I. (2007). The 'Hittite plague', an epidemic of tularemia and the first record of biological warfare. *Medical Hypotheses, 69*(6), 1371–1374. https://doi.org/10.1016/j.mehy.2007.03.012.

Tsoucalas, G., Laios, K., Karamanou, M., & Androutsos, G. (2014). Demystifying the epidemic among achaeans during the Trojan War. *Infezioni in Medicina, 22*(4), 342–348. http://www.infezmed.it/VisualizzaArticolo.aspx.

Tumpey, T. M., Basler, C. F., Aguilar, P. V., Zeng, H., Solórzano, A., Swayne, D. E., Cox, N. J., Katz, J. M., Taubenger, J. K., Pales, P., & García-Sastre, A. (2005). Characterization of the reconstructed 1918 Spanish influenza pandemic virus. *Science (New York, N.Y.), 310*(5745), 77–80. https://doi.org/10.1126/science.1119392.

Vitiello, L., Ilari, S., Sansone, L., Belli, M., Cristina, M., Marcolongo, F., Tomino, C., Gatta, L., Mollace, V., Bonassi, S., Muscoli, C., & Russo, P. (2022). Preventive measures against pandemics from the beginning of civilization to nowadays—How everything has remained the same over the millennia. *Journal of Clinical Medicine, 11*(7), 1960. https://doi.org/10.3390/jcm11071960.

Westra, P., Van Thiel, M. J. S., Vermeer, G. A., Soeterbroek, A. M., Scaf, A. H. J., & Claessens, H. A. (1986). Pharmacokinetics of galanthamine (a long-acting anticholinesterase drug) in anaesthetized patients. *British Journal of Anaesthesia, 58*(11), 1303–1307. https://doi.org/10.1093/bja/58.11.1303.

West, M. L. (2011). The Homeric question today. *Proceedings of the American Philosophical Society, 155*(4), 383–393.

WION. (2023). First H3N8 bird flu death reported in China.

Wong, D., Wild, M. A., Walburger, M. A., Higgins, C. L., Callahan, M., Czarnecki, L. A., Lawaczeck, E. W., Levy, C. E., Patterson, J. G., Sunenshine, R., Adem, P., Paddock, C. D., Zaki, S. R., Petersen, J. M., Schriefer, M. E., Eisen, R. J., Gage, K. L., Griffith, K. S., Weber, I. B., ... Mead, P. S. (2009). Primary pneumonic plague contracted from a mountain lion carcass. *Clinical Infectious Diseases, 49*(3), e33. https://doi.org/10.1086/600818, http://cid.oxfordjournals.org/content/by/year.

An eight-generation genealogy of Homer's Trojans

18

Eric L. Altschuler[1,2], *and Mehmet Oguz Namal*[3]
[1]Essen Health Care, Bronx, NY, United States, [2]Kessler Institute for Rehabilitation, West Orange, NJ, United States, [3]Bern, Switzerland

In a fascinating short paper, Bazopoulou-Kyrkanidou (2007) has constructed two Trojan family trees sourced from Homer's *Iliad*: an eight-generation genealogy of the Trojan Royal family and a tree listing many of Trojan King Priam's numerous offspring. The paper has never been cited or discussed. I don't think it's that the paper has been unappreciated, rather I think it has not been noticed. This may be due to the paper being more like an episode of the PBS series *Finding Your Roots with Henry Gates, Jr.* rather than a DNA or even classical genetic study. Also, the first family tree (Fig. 18.1) which is derived from information in the *Iliad* Book 20, lines 199–241, has appeared elsewhere, for example ((Homer, 2014) p. lxix). Nevertheless, the trees in the paper have lessons about human nature and also may teach us something about Homer as well.

In Fig. 18.1, notice two things in particular:

1. Only eight generations separate Zeus (1) from Hector (8). So, taking 20 or even 25 years/generation, Hector—the "crown prince" as it were of Troy—is separated by < ≈200 years from the immortals. Such a short amount of time could be a reasonable timespan of the history of Troy shortened for dramatic effect ("TV time"). However, this timeframe strains credulity if the *Iliad* represented a realistic history of the city or certainly the Earth. Indeed, in contrast, the Biblical age for the Earth based on genealogies is more than an order of a magnitude larger and by observing animal generations Homer's listeners would have appreciated how short eight generations are.
2. Hector, Priam's eldest son, is a member of the first family of Troy. (See also Fig. 18.2.) As we see from Fig. 18.1 Hector's third cousin is Aeneas. Several comments (pointed out by Murray (Homer, 1999, footnote p. 37) made to Aeneas in the *Iliad* (see Table 18.1) show him as the representative of the "second family" of Troy and that there seems to have been more than some tension between the two leading families of Troy. Homer must at some point observed and been witness to such tensions between rivaling families and factions.

Fig. 18.2 shows the offspring of King Priam's named in the *Iliad*. (Overall Priam notes he had fifty sons (Book 24, lines 495–497): "Fifty I had, when the sons of the Achaeans came; nineteen were born to me of the self-same womb [that is, by Queen Hecuba mother of Hector and Paris], and the others women of the palace bare.")

Homer describes the forces fighting on the Trojan side—the Dardanians and others—not as allies of the Trojans but rather mercenaries for Priam. Indeed, it is said it cannot be understood what those fighters are saying because of the mash up of different languages (*Iliad*, Book 4, lines 433–438): But for the Trojans, even as ewes

Medicine in Homer. DOI: https://doi.org/10.1016/B978-0-443-32978-4.00002-X

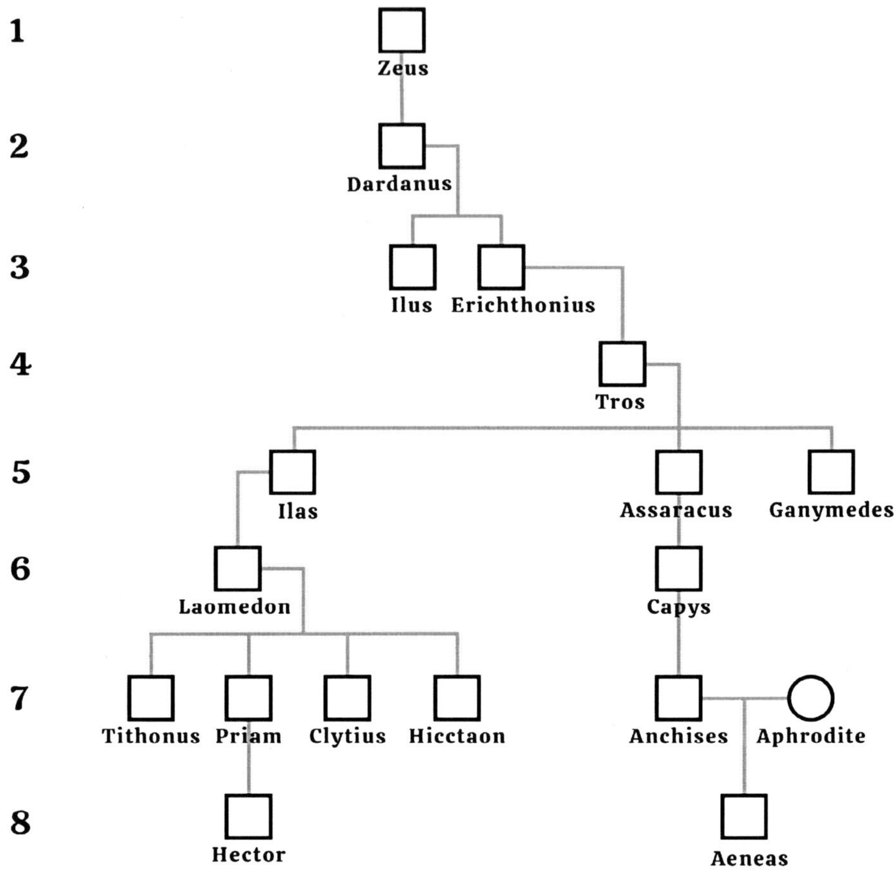

Figure 18.1 Eight-generation genealogy of Homer's Trojans.
Source: Author's (MON) own work.

stand in throngs past counting in the court of a man of much substance to be milked of their white milk, [435] and bleat without ceasing as they near the voices of their lambs: "even so arose the clamour of the Trojans throughout the wide host; for they had not all like speech or one language, but their tongues were mingled, and they were a folk summoned from many lands."

The Greeks, on the other hand, are described as a heroic allied coalition from various states under the command of Agamemnon (e.g., *Iliad*, Book 4, lines 427–432): that day did the battalions of the Danaans [Greeks] move, rank after rank, without cease, into battle; and each captain gave charge to his own men, and the rest marched on in silence; thou wouldst not have deemed [430] that they that followed in such multitudes had any voice in their breasts, all silent as they were through fear of their commanders; and on every man flashed the inlaid armor wherewith they went clad.

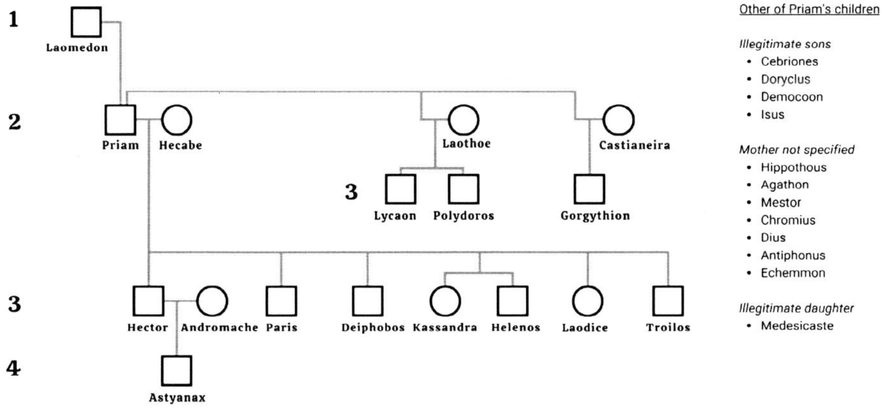

Figure 18.2 King Priam's family.
Source: Author's (MON) work adapted from Bazopoulou-Kyrkanidou (2007), used with permission.

Table 18.1 Comments made to Aeneas in the ***Iliad*** with relevance for his family as the "second family" of Troy.

Book, lines	Speaker	Comment made	Notes
13, 455–4622	Deïphobus (thoughts)	So spake he, and Deïphobus was divided in counsel, whether he should give ground and take to him as comrade some one of the great-souled Trojans, or should make trial by himself alone. And as he pondered this thing seemed to him the better—to go after Aeneas; and he found him standing last amid the throng, [460] forever was Aeneas wroth against goodly Priam, for that brave though he was amid warriors Priam honored him not a whit.	The Greeks were aware that Priam did not honor his cousin Aeneas.
20, 179–186	Achilles	then first unto Aeneas spake swift-footed goodly Achilles: "Aeneas, wherefore hast thou sallied thus far forth from the throng to stand and face me? Is it that thy heart biddeth thee fight with me [180] in hope that thou shalt be master of Priam's sovreignty amid the horse-taming Trojans? Nay, but though thou	Achilles is pointing out to Aeneas that whatever he does, and furthermore the risk such actions may entail not

(*Continued*)

Table 18.1 (Continued)

Book, lines	Speaker	Comment made	Notes
		slayest me, not for that shall Priam place his kingship in thy hands, for he hath sons, and withal is sound and nowise flighty of mind. [185] Or have the Trojans meted out for thee a demesne preeminent above all, a fair tract of orchard and of plough-land, that thou mayest possess it, if so be thou slayest me?	withstanding, he will not get a significant reward from Priam.
20, 306	Poseidon	For at length hath the son of Cronos come to hate the race of Priam; and now verily shall the mighty Aeneas be king among the Trojans, and his sons' sons that shall be born in days to come.	Poseidon is prophesying that Aeneas' will be king of the Trojans with generations of descendants.

Source: Quotes from Homer (1924). This work is licensed under a Creative Commons Attribution-ShareAlike 3.0 United States License.

Such, as always, the winning side in a war describes themselves as an alliance and the vanquished opponent as an evil coalition of fighters.

(Translations from Homer (1924): This work is licensed under a Creative Commons Attribution-ShareAlike 3.0 United States License.)

References

Bazopoulou-Kyrkanidou, E. (2007). Genetic drift. Descent, lineage, and pedigree of the Trojans in Homer's Iliad. *American Journal of Medical Genetics. Part A. 143A*(24), 2893–2895.

Homer (1924). *The Iliad.* Harvard University Press.

Homer (1999). *The Iliad: Volume II, Books 13-24.* Harvard University Press.

Homer (2014). In P. Jones (Ed.). *The Iliad.* Penguin Group.

Section 4

Further perspectives on medicine in homer

Can Homer's language and some mathematics help us to put a date on the Iliad?

19

Mark Pagel
University of Reading, Reading, United Kingdom

Homer's great story of the Trojan Wars as recounted in the *Iliad* tells us that the Homeric epics were almost certainly produced sometime after the 12th century BCE—if indeed the wars were ever fought. The poems attributed to Homer likely arose out of an oral tradition, only later being committed to print—but how much later? Herodotus—elevated by Cicero as "the father of history"—thought considerably later: writing in his Histories Book II.53 around 450 BCE, Herodotus held that Homer "lived, as I believe, not more than 400 years ago." Modern classicists largely agree: drawing on historical, literary, and archeological analyses, these scholars frequently assign a date to the *Iliad* of around the mid-8th century BCE (Fox, 2009; Janko, 1982), although some authors propose a more recent 7th century BCE date (West, 1966).

A different and complementary approach to answering this question is to ask whether, what we might call the "statistical-archaeology" of Homer's language itself, can along with some basic mathematics help us to put a date on when the oral tradition was finally written down.

Cognate words and rates of lexical change

Comparative linguists have known since at least Georgian times that languages predominantly evolve by a process of descent with modification from common or shared ancestral languages (Pagel, 2009). This process of descent means that words from different languages often display resemblances that betray their common origins. Such words, known as *cognates*, are not hard to find: for example, cognates meaning "water" exist in English (*water*), German (*wasser*), Swedish (*vatten*), and Gothic (*wato*), reflecting the descent of these words from what is thought to be the proto-Germanic word **watōr*. The Old English word *hund* is cognate to the modern English *hound*, and to the Germanic or Danish *hund* and the Dutch *hond*, again placing English among the Germanic languages. But English has now replaced *hound* for most uses with the newer word *dog* – an instance of what is known as *lexical replacement*.

These two phenomena—the retention of cognates and lexical replacement—could, if they show regularities in their occurrence, be used to supply dates as to when two languages might have separated from a common ancestral language in the past. If we

Medicine in Homer. DOI: https://doi.org/10.1016/B978-0-443-32978-4.00011-0

could identify a common set of meanings that all languages would be expected to have, such as the names of body parts, common physical features of the environment, relationship words like he, she, father, mother, and so on, we could write down the words for these meanings from related languages and assess them for cognancy versus lexical replacement. Then, if we knew the rates at which words tended to decay and get replaced by the newer noncognate words, we could use the number of cognates that pairs of languages had in common along with some straightforward mathematics to estimate when they last shared a common ancestor.

Both of these resources exist. Working in the 1950s the linguist Morris Swadesh produced a list of 100–200 common vocabulary words that he called the fundamental vocabulary (Swadesh, 1952), and which can be expected to be found in all languages. Around 60 years later, colleagues of mine and I were able to estimate rates of lexical replacement for each of the words in Swadesh's list (Pagel et al., 2007). A word's rate of lexical replacement measures the instantaneous rate at which a word such as *hund* gets replaced by a new unrelated or noncognate word such as *dog*. Most words have a 50% chance of being replaced by a new noncognate word every 2000 to 3000 years (a word's linguistic *half-life* (Pagel et al., 2007)), but among the words in the Swadesh list there is at least a 100-fold variation in rate of change. Some adjectives like *dirty* have linguistic half-lives under a millennium whereas others, such as the lower-limit number words (the words *one* to *five*) are expected to last over 10,000 years.

The variation in rates of lexical replacements makes these vocabulary items promising candidates for estimating the time of divergence between pairs of languages: rapidly evolving words might be expected to change even among relatively closely related languages, whereas more slowly evolving words might remain cognate even among distantly related languages—if either all or none of the words changed in a comparison of two languages, they would not be useful for establishing a date. More generally, given a set of rates and a set of vocabulary items that have either remained cognate or not between pairs of languages, we can ask what divergence times best satisfy the observed distribution of cognancy relations among those languages.

A date for the Homeric epics

With colleagues, I have applied that logic to estimate the age of the Homeric epics, using vocabulary items from Swadesh's fundamental vocabulary (Swadesh, 1952). We studied Swadesh's fundamental vocabulary items for three languages: Homeric and Modern Greek, and Hittite. We can think of Homeric Greek as a direct ancestor to Modern Greek Fig. 19.1. Hittite is an extinct language in the Anatolian branch of the family of languages of which Greek, English, the Romance, Slavic, and Indo-Iranian languages are a part and known as Indo-European. As such Hittite can be thought of as a "sister" language to Homeric and Modern Greek, last sharing a common ancestor with those two languages sometime around the origin of the Indo-European language family.

Preserved cuneiform scripts dating to the 13th to 16th centuries BCE provide a source of Hittite's vocabulary and age (Altschuler et al., 2013; references therein).

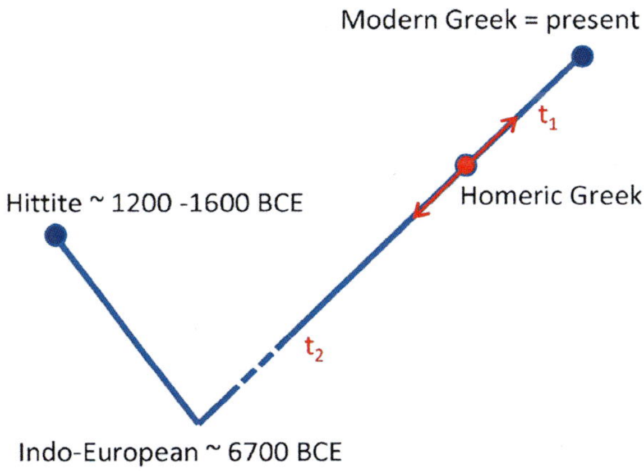

Figure 19.1 Evolutionary (phylogenetic) relationships among Hittite, Homeric Greek, and Modern Greek with approximate ages shown (note: lines not to scale, age for Indo-European from Gray and Atkinson (2003).

Other work (Gray & Atkinson, 2003) suggests an age for the Indo-European language family of around 6700 years BCE. Modern Greek is dated to the present, and Herodotus' remark can be used to restrict the age of Homeric Greek to be no younger than 450 BCE when he lived. Together with these dates, comparisons of the number of shared cognates among the three pairs of languages – Hittite, Homeric, and Modern Greek can be used to "bracket" Homeric Greek, providing overlapping evidence to where in time Homer's works fall between them (branch denoted "t_1" in Fig. 19.1).

We were able to extract information about shared cognates among the three language pairs for 173 of the 200 Swadesh fundamental vocabulary items (Altschuler et al., 2013). As expected, Hittite and Modern Greek share the fewest cognates (13.3%), followed by Hittite and Homeric Greek at 19.1% of their vocabulary. By comparison, the same figure for Homeric and Modern Greek is 50% (Table 19.1), suggesting, as we would expect, that they are more closely related to each other than either is to Hittite. In each of these three pairwise comparisons the words remaining cognate between a pair have slower rates of lexical replacement than those that have changed (Table 19.1).

This analysis gives us a rank ordering for the times separating any two pairs of languages but doesn't give us estimated dates. But here we can use some simple mathematics to estimate those dates. If we knew, for example, that a word was replaced by a new noncognate word every 1000 years on average, and we observed that two languages differed on that word, we could say with 50% confidence that they had been separated for around 693 years. This is the linguistic half-life for a word that gets replaced on average once every 1000 years—that is, 693 years is the time we would have to wait to be 50% sure that a word that "decayed" at a rate of once per thousand years would get replaced.

Table 19.1 Comparisons among Hittite and Homeric and Modern Greek.

Data	Hittite—Homer	Hittite—Modern Greek	Homer—Modern Greek	Notes
Cognate/not cognate[1]	33/140	23/150	87/86	[1]Refers to number of words out of 173 (see text) judged cognate between the two languages.
Percent cognate	19.1	13.3	50.3	
Replacement rate for words that remained cognate[2]	1.62 ± 1.26 (sd)	1.28 ± 1.03	1.19 ± 1.63	[2]Rates of lexical replacement taken from Pagel et al. (2007).
Replacement rate for noncognate words	3.32 ± 1.77	3.26 ± 1.76	3.54 ± 1.83	

Given a set of words and their rates of change, we can use this logic to estimate the probable time separating two languages. That estimate is the single guess that best satisfies—in our case—the set of 173 judgments of cognate versus noncognate that we can make between any pair of languages along with their rates of lexical replacement.

The analysis we conducted (Altschuler et al., 2013) returns an average estimate of the date for Homer's works of 762 BCE with 95% confidence intervals from 450 to 1157 BCE. It is intriguing in this light that the "Nestor's cup" (Nestor's Cup Pithekoussai, 2024), a vase excavated from an Ancient Greek site in Italy, contains an inscription that some scholars think refers to a line from the *Iliad* and is dated to ~740–720 BCE. Equally, the 95% upper (older) limit falls in the middle of the 12th century, the period during which some scholars think the Trojan wars might have been fought.

Conclusions

The 762 BCE date for the Homeric epics, derived from vocabulary data, agrees with the commonly accepted 8th century BCE origin of these great works based on archaeological, literary, and historical information. It is consistent with Herodotus' remark that would place the epics around 850 BCE but would treat as unlikely any suggestion that Homer might have been a "war correspondent" recording the events of the Trojan War as they happened, if indeed they ever did (of course, the epics were originally an oral tradition so we cannot know if our dates reflect when they were produced or when they were eventually written down).

Our estimated date falls roughly in the middle of the classicists' and historians' preferred date for Homer, representing a prediction spanning nearly three millennia. This, along with the consistency of the results (Table 19.1), demonstrates a remarkable regularity in the ways that words are replaced over time and illustrates that language can be used, like genes in biology, to aid in the investigation of questions in history, archeology, and anthropology (Barbrook et al., 1998). Along with the historical and other accounts of Homer's great works, our analysis of common vocabulary items in the *Iliad* increases our confidence in its age and shows how even what were originally oral traditions can preserve traces of history.

Acknowledgments

I thank Theo van den Hout for help in identifying cognate words between Hittite and Homeric Greek and Hittite and Modern Greek and the European Research Council for its generous support of this research.

References

Altschuler, E. L., Calude, A. S., Meade, A., & Pagel, M. (2013). Linguistic evidence supports date for Homeric epics. *BioEssays, 35*(5), 417–420. https://doi.org/10.1002/bies.201200165.

Barbrook, A. C., Howe, C. J., Blake, N., & Robinson, P. (1998). The phylogeny of The Canterbury Tales. *Nature, 394*(6696), 839. https://doi.org/10.1038/29667, http://www.nature.com/nature/index.html.

Fox, R. L. (2009). *Travelling Heroes: Greeks and their myths in the epic age of Homer*. Random House, 360–364.

Gray, R. D., & Atkinson, Q. D. (2003). Language-tree divergence times support the Anatolian theory of Indo-European origin. *Nature, 426*(6965), 435–439.

Janko, H. (1982). *Hesiod and the Hymns: Diachronic development in epic diction*. Cambridge University Press.

Nestor's Cup (Pithekoussai). (2024). https://en.wikipedia.org/wiki/Nestor%27s_Cup_(Pithekoussai).

Pagel, M., Atkinson, Q. D., & Meade, A. (2007). Frequency of word-use predicts rates of lexical evolution throughout Indo-European history. *Nature, 449*(7163), 717–720.

Pagel, M. (2009). Human language as a culturally transmitted replicator. *Nature Reviews. Genetics, 10*(6), 405–415. https://doi.org/10.1038/nrg2560.

Swadesh, M. (1952). Lexico-statistic dating of prehistoric ethnic contacts. In *Proceedings of the American Philosophical Society*, *96*, 453–463.

West, M. L. (1966). *Hesiod: Theogony*. Clarendon Press.

Analyzing the social networks of Homer's *Iliad* and *Odyssey*

20

Pádraig MacCarron
Department of Mathematics & Statistics, University of Limerick, Limerick, Ireland

Introduction

A new approach to studying ancient literature is that of *complex networks*. A network is a collection of objects that have interactions between them. Although a simple concept, many mathematical tools have been developed to analyze, compare, and model the structure of networks. Networks have a wide range of applications, from archeology (e.g., looking at maritime trade between sites in the southern Aegean in the Bronze Age; Knappett et al., 2008) to zoology (e.g., studying the relationships between primates in a group (Mac Carron & Dunbar, 2016)). They have also been used in the analysis of medieval history (Gramsch et al., 2016) and Greek and Roman mythology (Choi & Kim, 2007).

A *social network* is a particular type of network where the objects are represented by people, and the interactions represent some kind of relationship, usually friendship. Social networks have developed ideas such as "the friendship paradox," the notion that, on average, your friends have more friends than you, and to understand how many friends one can have (Dunbar, 2012). They have many applications, for example, modeling the spread of disease or information (O'Sullivan et al., 2015) or maximizing the efficiency of vaccination-deployment schemes (Miller & Hyman, 2007).

In my own research, we have quantitatively compared the social networks of epics and myths from different cultures (e.g., the *Iliad*, the Anglo-Saxon *Beowulf*, and the Irish *Táin Bó Cúailnge*) (MacCarron & Kenna, 2012), as well as studied the social networks of the same culture (e.g., sagas of the Icelanders) (MacCarron & Kenna, 2013). Within this, we observed the networks of two sagas, *Njáls saga* and *Laxdœla saga*, were found to be difficult to separate the networks after merging them. This supported the hypothesis that the author of *Njáls saga* had used *Laxdœla saga* as a source. A similar question was asked in Yose et al. (2017) regarding the *Poems of Ossian*. These poems were purported to have been found by James Macpherson in the Scottish highlands. They were published and highly regarded, as well as frequently compared to the works of Homer. However, this sparked a major literary controversy as Macpherson never produced his original sources and the work was highly similar to material based on the Irish Fionn Mac Cumhaill and his son Oisín. Comparing the networks of Macpherson's work to the Irish Fenian cycle and the works of Homer, we observed the structure of the network was highly similar to that of the Irish texts and quite unlike the networks of the *Odyssey* or *Iliad*. More recently, this network application to mythology is tackling questions such as the effect of information loss

Medicine in Homer. DOI: https://doi.org/10.1016/B978-0-443-32978-4.00015-8

on one culture's myths as well as quantifying the role of major characters in the full mythology (MacCarron, 2024).

In this chapter, we construct and compare the social networks of the *Iliad* and the *Odyssey*. We create the social networks of both positive and negative interactions for each of these and analyze their network properties. As there are many shared characters between the two, we can merge these into a single network. We will use *community detection algorithms* on this network and see if we can separate the two epics in the same way as the sagas of the Icelanders in MacCarron and Kenna (2013).

We then compare these to Virgil's *Aeneid*, which is heavily inspired by the works of Homer. The *Aeneid* is an epic Latin poem composed by Virgil between 29 and 19 BC. The story deals with the protagonist Aeneas and his destiny to begin the Roman empire. After the events of the Trojan War, a group of Trojans led by Aeneas escape. They land in Carthage where the queen, Dido, falls in love with Aeneas (with the help of the gods). After a year they leave and eventually settle in Latium. Some of the locals, led by Turnus, engage them in a series of battles. The story culminates with Aeneas defeating Turnus in single combat. Many scholars have discussed the influence of Homer's work on the *Aeneid*, as well as Virgil's intent for this to be recognized and to be familiar to the works of Homer. Clausen (1964) describes the *Aeneid* as prolonged literary allusion to Homer, and Knauer and Nicolaus (1964) discuss how the first half mirrors the *Odyssey* and the section in Latium mirrors the *Iliad*.

Finally we analyze the roles of the physicians in the combined network of these two of Homer's epics. Here we focus on some individual properties to assess how central or important these characters are in Homer's work which ties in with the theme of this volume. In the next section, the methods and technical details are explained, Section 20.3 presents the results of the analyses, and the final section summarizes and contextualizes these results.

Methods

A *graph*, or *network*, is a mathematical object comprising *nodes* (or vertices) and *edges* (or links). Nodes are connected to each other by edges. The study of graph theory began in the 18th century but the earliest known application to social systems is from the 1930s showing friendships between school children (Moreno & Levy, 1934). In social networks, a node is often referred to as an actor or an ego (and their neighbors are often called alters).

A simple network is shown in Fig. 20.1 with N=20 nodes connected with a total of L=37 edges. The number of neighbors a node has is called its *degree* k. For example, the node A has a degree of k=5 and node B has k=6. The mean degree here is $\langle k \rangle$=3.7, indicating each node is connected, on average, to between three and four others.

Initially, graphs where nodes were connected at random were modeled to represent social networks (the most common type being Erdős–Rényi random graphs). With access to large datasets from online sources in the 1990s, and increased computational power, it was observed that real-world networks such as a network of movie

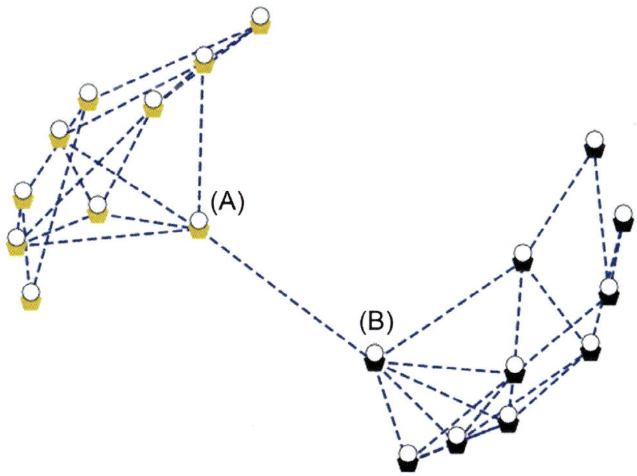

Figure 20.1 An example of a social network with two communities. Nodes A and B have the highest betweenness centralities as they are on all paths between nodes in the opposing community.

actors connected by working on the same film, or the US power-grid network, had structural properties that were different to model graphs (Watts & Strogatz, 1998). This gave rise to the field which is now known as *complex networks*.

In a complex network, the histogram of the degree of nodes is right-skewed, unlike a random graph which is binomially distributed. This informs us that, in a complex network, the average number of neighbors is not representative of the data, some nodes will have a degree orders of magnitude larger than the mean. Fig. 20.2 shows the degree distributions for an Erdős–Rényi random graph on the left and a model complex network using the Barabási-Albert model on the right (Barabási & Albert, 1999). Both of these networks are constructed to have the same number of nodes $N=1,000$ and mean degree $\langle k \rangle=50$. The random graph on the left in Fig. 20.2(A) is symmetric about the mean with the degree of most nodes within less than two standard deviations. The complex network in Fig. 20.2(B) however is right-skewed meaning most nodes have a degree much lower than the mean, with a small number far larger. The distributions are shown as probability density functions, however later in the chapter we will instead use the complementary cumulative frequency distribution. This is a transformation that is useful for visualization as it reduces the noise in the tail. In Fig. 20.2(B), we observe many instances of just one node having a specific degree k. By taking the cumulative frequency instead, this changes to a node having at least that value, so the probability always decreases beginning at one on the left.

Complex networks also tend to have nodes in highly clustered modules, and, on average, a small number of steps separating two nodes (a property known as *small world*). Social networks are found to be complex with all these properties. Unlike most other complex networks, however, social networks are often found to be

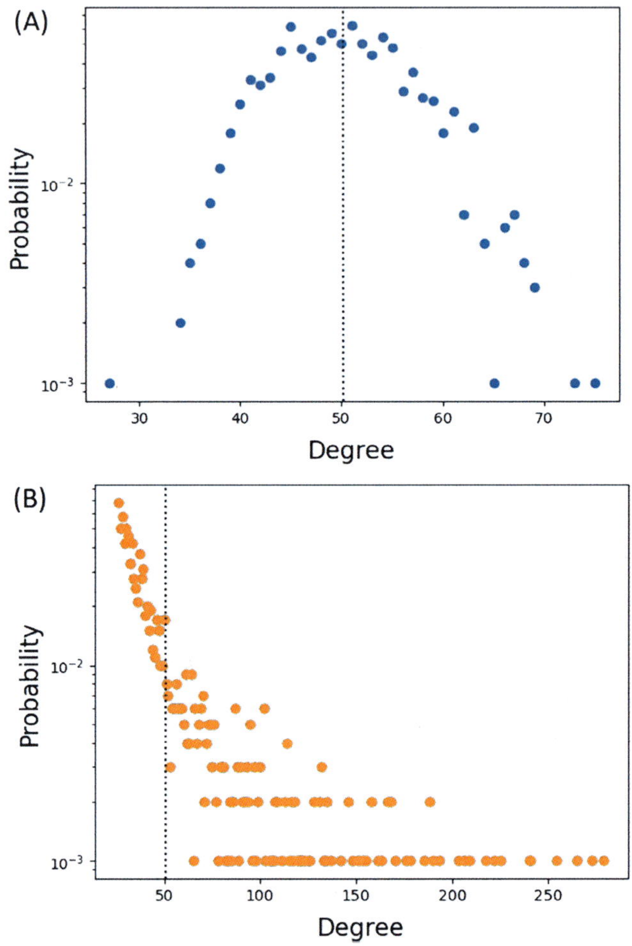

Figure 20.2 Model degree distributions. (A) The degree distribution for an Erdős–Rényi random graph constructed with 1000 nodes and an average degree of 50. This is binomially distributed (can be approximated as a Poisson distribution for large graphs). (B) The degree distribution for a model complex network. This network is the same size and has the same average degree as the random graph in panel (A), but the shape of the distribution is instead right-skewed with most nodes having a degree lower than the mean and a few with significantly higher degrees.

assortative (Newman & Park, 2003), meaning that nodes with similar properties tend to connect to one another (related to the notion of *homophily* in sociology). The degree assortativity r is measured with a Pearson correlation coefficient on the degrees of all neighboring nodes. If this is positive, $r > 0$, it is said to be assortative and negative, $r < 0$, disassortative. The network in Fig. 20.1 is assortative with r=0.41. This tells us that nodes generally connect to nodes with a similar number of neighbors. A star-shaped network, with all but one node connected to a central node (which is thus connected to every other nodes) is totally disassortative with r=–1.

As well as these structural properties of the network, we are often interested in individual node properties. These are often called *centrality* measures. The most common is the aforementioned degree (the number of neighbors a node has). However, sometimes this is not the most useful property to identify an important node. Another core centrality measure is called the *betweenness centrality*. This finds the number of shortest paths between each pair of nodes and ranks nodes based on how many of those paths they are on. For example, in Fig. 20.1, if any information was to pass from a node in the light-coloured community on the left to a node in the dark community on the right, that information would have to pass through nodes *A* and *B*. Hence these two nodes have a higher betweenness centrality than other nodes with the same degree.

Another feature of complex networks is that of community structure. A *community* in a network is a densely clustered group of nodes. The network presented in Fig. 20.1 shows two communities, the darker nodes on the lower right are one community, and the lighter ones on the upper left the second. Many methods have been developed to attempt to find communities in large networks. The method used here to detect communities is the Girvan-Newman method (Girvan & Newman, 2002). This works by recursively removing edges with the highest betweenness centrality. In Fig. 20.1, for example, removing the edge connecting node *A* and node *B*, will separate the two communities. This method will be applied to the combined network of the *Iliad* and the *Odyssey* to determine the major communities present.

Here, the texts used are Homer's *Iliad* (translated by Rieu (2003)) and *Odyssey* (translated by Kirk (1980)), and Virgil's *Aeneid* (translated by Oakley (2004)). To gather the data, each narrative is closely read, and any time two characters interact, or it is clear they have interacted in the past, an edge is recorded. We split the interactions into negative –when they interact in combat– and positive –when they interact in any other way. We *weigh* the edges based on how many chapters these interactions appear in and so this represents the strength of their relationship. However, in this chapter, we do not use these weights as they do not significantly affect any of the results presented below. They are useful when analyzing the change of the network as the story evolves however like in Gessey-Jones et al. (2020), for example.

Results

We begin by comparing the structural properties of three individual networks, as well as merging the two Homer epics for which we will refer to as the amalgamated-Homer network. Table 20.1 shows the size N, number of edges L, mean degree $\langle k \rangle$, the mean path length $\langle \ell \rangle$, the mean path length of the same sized Erdős–Rényi random graph $\langle \ell_r \rangle$, the assortativity r, and the assortativity for positive interactions r_+ for each of the epics and the amalgamated Homer network.

The *Iliad* is the largest of the individual epics with Virgil's *Aeneid* between the two Homeric epics in size. However, the *Aeneid* has less interactions than the *Odyssey*

Table 20.1 The size N, number of edges L, mean degree $\langle k \rangle$, the mean path length $\langle \ell \rangle$, the mean path length of the same sized Erdős–Rényi random graph $\langle \ell_r \rangle$, the assortativity r, and the assortativity for positive interactions r_+ for the two Homer epics, the *Aeneid*, and the amalgamation of the *Iliad* and *Odyssey*.

Network	N	L	$\langle k \rangle$	$\langle \ell \rangle$	$\langle \ell_r \rangle$	r	r_+
Iliad	694	2686	7.74	3.49	3.42	–0.08	0.09
Odyssey	301	1020	6.78	3.29	3.18	–0.08	–0.07
Aeneid	444	985	4.44	3.57	4.20	–0.13	–0.06
Iliad + Odyssey	874	3611	8.26	3.39	3.43	–0.09	–0.01

giving characters a smaller average degree. When the mean path length $\langle \ell \rangle$ of a network is similar in size to that of a random graph of the same size $\langle \ell_r \rangle$, it is said to be a small world. This is true for both the Homeric epics but not the *Aeneid*.

Many social networks are found to be assortative, $r > 0$, meaning that nodes tend to associate with nodes of a similar degree (Newman & Park, 2003). Those social networks do not have data on hostilities, therefore when we compare these, we focus only on the non-combat interactions r_+. Here we see that only the *Iliad* has a positive value indicating that it is assortative, making it the most similar to real-world networks (Newman & Park, 2003) and having similar network properties to some sagas of the Icelanders (MacCarron & Kenna, 2013).

Merging the social networks of Homer, we see that they have over 100 shared characters. The merged network properties are closer to that of the larger *Iliad* than the *Odyssey*. Fig. 20.3(A) shows the complementary cumulative form for the degree distributions for the amalgamated network and the *Iliad*. We see that the shapes of the distributions are similar; however, Odysseus' degree becomes much larger (as seen by the outlier on the right). The largest degree in the *Iliad* is Hector with k=106. Odysseus is fifth highest with k=61. In the *Odyssey*, however, his degree is k=112 and, in the amalgamated Homer network, this increases to k=157 (for a list of the top 10 characters by degree see Table 20.2). A jump in the degree distribution can be observed in Fig. 20.3(A) around k=33. This disjoint is due to a group of Nereids (sea nymphs) who all appear to comfort Achilles' mother, Thetis. It is unusual in these networks to have a large cluster of characters who mostly do not interact with any other characters in the entire story.

Fig. 20.3(B) shows the degree distributions for the *Odyssey* and the *Aeneid*. Here we observe that their distributions are quite similar. Using the methods of maximum likelihoods to fit the distributions (Mannion & MacCarron, 2023), we find that they are both well fitted by Weibull distributions. However, it is clear in both cases that two characters fall far to the right of the rest of the distribution. These two characters are Odysseus and his son Telemachus in the *Odyssey*, and Aeneas and Turnus in the *Aeneid*. If we fit the distributions without the top two nodes, the rest of the network are well approximated by exponential distributions, though with different scaling parameters.

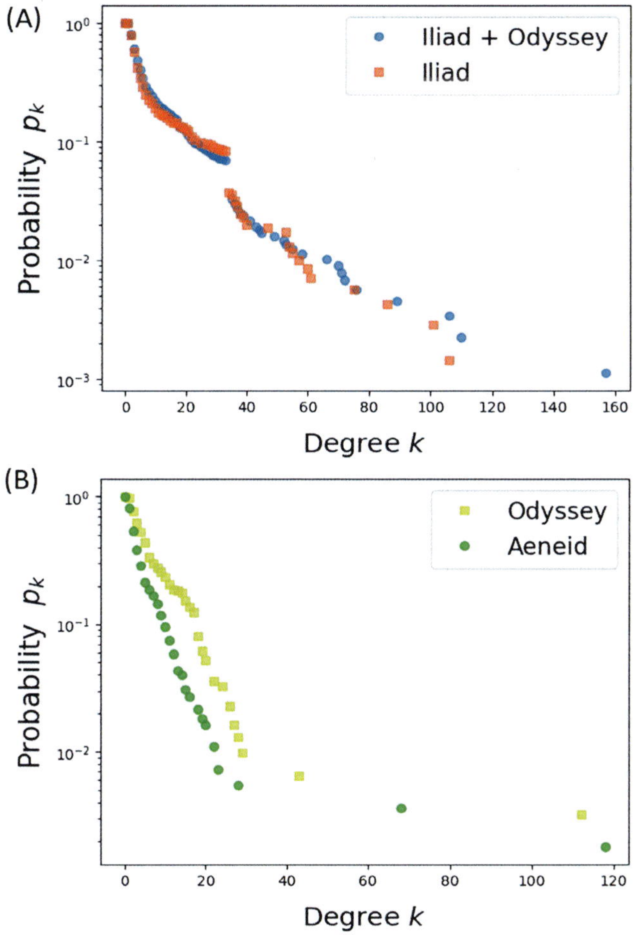

Figure 20.3 The degree distributions for each of the networks. (A) On the left shows the degree distribution for the amalgamated *Iliad* and *Odyssey* networks (circles) and the *Iliad* on its own (squares). (B) The network-structural similarity of the *Odyssey* and *Aeneid* with their similar degree distribution shapes.
Source: Author's own work.

This apparent structural similarity between Virgil's *Aeneid* and the *Odyssey* is not particularly surprising given that the first half of the *Aenied* in particular is often compared to the *Odyssey* and Virgil's apparent desire to rival the works of Homer (see, e.g., Clausen, 1964). However, it is worth noting that while they have some structural similarities, the *Aeneid* is the only one of these networks that is not small-world, making it more different to real-world social networks than the others.

We next look at performing community detection on both the *Iliad* and the amalgamated network. As mentioned when discussing the *Iliad*'s degree distribution

Table 20.2 Most central characters.

Rank	Degree (*k*)	Betweenness centrality (*k*)
1	Odysseus (157)	Odysseus (157)
2	Agamemnon (110)	Agamemnon (110)
3	Hector (106)	Achilles (89)
4	Achilles (89)	Patroclus (66)
5	Priam (76)	Diomedes (58)
6	Menelaus (72)	Hector (106)
7	Nestor (71)	Zeus (70)
8	Zeus (70)	Nestor (71)
9	Patroclus (66)	Helen (33)
10	Diomedes (58)	Hermes (26)

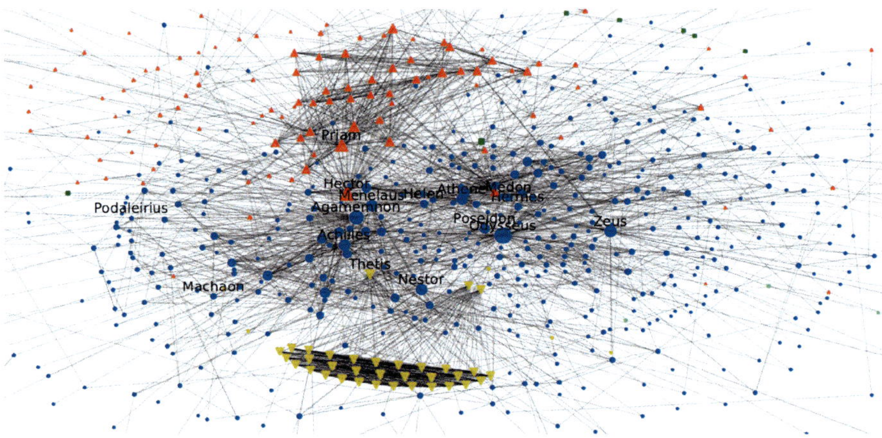

Figure 20.4 Homer's network. The network of the *Iliad* and the *Odyssey*. Nodes are shaped and colored based on the Girvan-Newman community detection algorithm with the Trojans as the triangles at the top, the Nereids as the inverted triangles at the bottom, and the remainder (mostly Greek characters) as circles. The 15 most central characters are named (as well as two physicians).

above, the group of Nereids, with the exception of Thetis, do not interact with the rest of the characters in the *Iliad*. As a result, when performing community detection, they are found as one cluster. Using either the *Iliad* or the amalgamated Homer network, the community detection method finds three large communities. These are shown in Fig. 20.4. The inverted triangles at the bottom or the Nereids. The triangles at the top are mostly Trojan characters, and the circular nodes are mostly Greek characters. Therefore even when merging the *Iliad* and the *Odyssey*, the Trojans are easier to separate out from a social-network perspective, than splitting the two narratives. This shows how strongly connected these two epics are.

Most central characters

The top 15 characters in the combined Homer network in order of betweenness centrality are named in Fig. 20.4 (as well as two physicians, see below). Some of these are Olympians such as Zeus, Athena, and Poseidon. They tend to end up in the community with the faction they support in the Trojan War.

Table 20.2 shows the top 10 most central characters ordered by the degree first and then betweenness for the amalgamated Homer network (each character's number of neighbors is given in parenthesis after). Unsurprisingly Odysseus is the most central by both measures as he is a major character in the *Iliad* and the protagonist of the *Odyssey*. The Greek king Agamemnon also features in both and is ranked second in both measures. Many of the same characters appear highly in both measures; however, the major differences are the addition of Helen and the god Hermes in the betweenness centrality top ten. Helen, for whom the Trojan War was essentially waged, has the 30th highest degree in the amalgamated network but is the ninth most important when ordering by betweenness centrality.

As per the theme of this volume, we next turn our attention to some of the physicians in Homer's epics. Machaon and Polidarius (spelled Podaleirius in the translation by E.V. Rieu) are two important physicians of the Greeks (Hajar, 2002). We first look at their own centrality in the network, followed by their centrality in the Greek community (as found by the community detection algorithm). These characters are both labeled in Fig. 20.3 appearing on the left-hand side of the network. Neither of these are highly central in the network, looking at the full network Machaon is ranked 187 of the 881 characters in terms of betweenness centrality. If we use the Greek community as defined by the community detection algorithm, of these 523 characters, in terms of degree, Machaon ranks 84th. Polidarius is less important for the social network ranking 489 by betweenness in the full network. When looking at the importance of just the Greek community by degree, he ranks in the top half at 231.

Characters such as Patroclus and Sthenelus (a friend of Diomedes) can be considered as "lay physicians" in the *Iliad*. Both of these have high betweenness centralities in the amalgamated network and the *Iliad* showing their overall importance. In the amalgamated network, Patroclus is the fourth most important and Sthenelus the 58th. Focusing just on the *Iliad* these increase to third and 49th.

Conclusions

Homer's epics have influenced many authors and story-tellers over the centuries. One particularly early example of this is the poet Virgil and his *Aeneid* which follows on directly from the siege of Troy. When comparing the social networks of the *Aeneid* and that of the *Odyssey* and the *Iliad*, we observe that it has some similar structural properties to the *Odyssey* but distinct from the *Iliad*. This is most apparent from the degree distribution as shown in Fig. 20.3(B) and almost identical values of assortativity. However, the *Aeneid* is the only network here not to have the small-world property making it less similar to real-world social networks. More network metrics for these and other mythological and epic networks can be found and compared in MacCarron (2014). There it can be seen that it is rare for these narrative networks to not be small-world (unless they contain large genealogies which create long chains in the network).

A similar comparison of determining social network similarities was performed in Yose et al. (2017). There, the idea was to determine if the network of the *Poems of Ossian* was more similar to the Irish material it was argued to be based on, or the works of Homer to which it was frequently compared. This was found to have highly similar structural properties to the Irish network, perhaps as a sign of it being a forgery and using those sources but trying to hide it. In the case of Virgil however, he is not trying to hide his influences, but they are just that: influences and inspiration. As a result, the network properties are distinguishable, as he has constructed a new story based on preexisting material. This is more similar to the suggestion that the author of the Icelandic *Njáls saga* had used *Laxdœla saga* as a source (MacCarron & Kenna, 2013). Hence the network properties are similar, but still distinguishable unlike the case of Macpherson's *Ossian*.

As hypothesized in MacCarron (2014), the social network properties are unlikely to be related to cultural signatures, but more indicative of the type of story. For example, both the *Odyssey* and the *Aeneid* have significant portions where the protagonist is moving around different locations and interacting with different communities and creatures on their journey. This is similar to *Beowulf* and the Icelandic *Gísla saga*. As a result, all of these have highly similar network properties. Contrast this with epics like the *Iliad*, this is mostly set in a single location and even takes place over a short period of time (a few weeks). The social network properties here are similar to the Icelandic *Laxdœla saga* which deals with people in one region. This is likely another reason why Virgil's *Aeneid* has different network properties.

The *Iliad* itself is one of a few narrative networks that has properties similar to real-word networks, particularly in terms of assortativity, a property used to demonstrate the artificiality of the Marvel universe (Gleiser, 2007). Although tempting to argue this is due to the realistic social world Homer creates, the previous point also stands, that the narrative allows us to get familiar with two large communities: the Greeks and Trojans. So well realized are these communities, that the network still separates them even when adding all the additional characters of the *Odyssey* in there. In the *Odyssey*, with the exception of those at Ithica, most of the communities Odysseus encounters are not as well realized as those in the *Iliad*. Nevertheless, it is compelling that the *Iliad* has these realistic social-network properties given archeological evidence for a Trojan War (Korfmann, 2004). From a psychological perspective, some scholars argue that for a narrative to be successful, the social world they create must be believable, a notion known as the principle of minimal difference (Palmer, 2010). Therefore this might help explain why a story like the *Iliad* has endured for so many centuries in popularity and interest.

References

Barabási, A.-L., & Albert, R. (1999). Emergence of scaling in random networks. *Science (New York, N.Y.)*, *286*(5439), 509–512. https://doi.org/10.1126/science.286.5439.509.

Choi, Y.-M., & Kim, H.-J. (2007). A directed network of Greek and Roman mythology. *Physica A: Statistical Mechanics and its Applications, 382*(2), 665–671. https://doi.org/10.1016/j.physa.2007.04.035.

Clausen, W. (1964). An Interpretation of the Aeneid. *Harvard Studies in Classical Philology, 68*, 139. https://doi.org/10.2307/310802.

Dunbar, R. I. M. (2012). Social cognition on the Internet: Testing constraints on social network size. *Philosophical Transactions of the Royal Society B: Biological Sciences, 367*(1599), 2192–2201. https://doi.org/10.1098/rstb.2012.0121.

Gessey-Jones, T., Connaughton, C., Dunbar, R., Kenna, R., MacCarron, P., O'Conchobhair, C., & Yose, J. (2020). Narrative structure of A Song of Ice and Fire creates a fictional world with realistic measures of social complexity. *Proceedings of the National Academy of Sciences of the United States of America, 117*(46), 28582–28588. https://doi.org/10.1073/pnas.2006465117.

Girvan, M., & Newman, M. E. J. (2002). Community structure in social and biological networks. *Proceedings of the National Academy of Sciences of the United States of America, 99*(12), 7821–7826. https://doi.org/10.1073/pnas.122653799.

Gleiser, P. M. (2007). How to become a superhero. *Journal of Statistical Mechanics: Theory and Experiment, 2007*(09), P09020. https://doi.org/10.1088/1742-5468/2007/09/p09020.

Gramsch, R., MacCarron, M., MacCarron, P., & Yose, J. (2016). *Medieval historical, hagiographical and biographical networks*. Springer Science and Business Media LLC, 45–69. http://doi.org/10.1007/978-3-319-39445-9_4.

Hajar, R. (2002). Learning ancient Greek medicine from Homer. *Heart Views, 3*.

Kirk, G. S. (1980). *The Odyssey*. Oxford University Press.

Knappett, C., Evans, T., & Rivers, R. (2008). Modelling maritime interaction in the Aegean Bronze Age. *Antiquity, 82*(318), 1009–1024. https://doi.org/10.1017/s0003598x0009774x.

Knauer, G., & Nicolaus (1964). Vergil's Aeneid and Homer. *Greek, Roman and Byzantine Studies, 5*, 61–84.

Korfmann, M. (2004). Was there a Trojan War? *Archaeology, 57*, 37–41.

Mac Carron, P., & Dunbar, R. I. M. (2016). Identifying natural grouping structure in gelada baboons: A network approach. *Animal Behaviour, 114*, 119–128. https://doi.org/10.1016/j.anbehav.2016.01.026.

MacCarron, P. (2024). Quantifying the role of supernatural entities and the effect of missing data in Irish sagas. *Condensed Matter Physics, 27*(3), 33803.

MacCarron, P. (2014). *A network theoretic approach to comparative mythology*.

MacCarron, P., & Kenna, R. (2013). Network analysis of the Íslendinga sögur – The Sagas of Icelanders. *The European Physical Journal B, 86*, 1–9.

MacCarron, P., & Kenna, R. (2012). Universal properties of mythological networks. *EPL (Europhysics Letters), 99*(2), 28002. https://doi.org/10.1209/0295-5075/99/28002.

Mannion, S., & MacCarron, P. (2023). A robust method for fitting degree distributions of complex networks. *Journal of Complex Networks, 11*(4). https://doi.org/10.1093/comnet/cnad023.

Miller, J. C., & Hyman, J. M. (2007). Effective vaccination strategies for realistic social networks. *Physica A: Statistical Mechanics and its Applications, 386*(2), 780–785. https://doi.org/10.1016/j.physa.2007.08.054.

Moreno, J., & Levy (1934). *Who shall survive? A new approach to the problem of human interrelations*.

Newman, M. E. J., & Park, J. (2003). Why social networks are different from other types of networks. *Physical Review E, 68*(3). https://doi.org/10.1103/physreve.68.036122.

Oakley, M. (2004). *The Aeneid*. Wordsworth Classics.

O'Sullivan, D. J. P., O'Keeffe, G. J., Fennell, P. G., & Gleeson, J. P. (2015). Mathematical modeling of complex contagion on clustered networks. *Frontiers in Physics, 3*. https://doi.org/10.3389/fphy.2015.00071.

Palmer, A. (2010). *Storyworlds and groups. Introduction to cognitive cultural studies*. The Johns Hopkins University Press, 176–192. http://muse.jhu.edu/books/9781421400280.

Rieu, E. V. (2003). *The Iliad*. Penguin Classics.

Watts, D. J., & Strogatz, S. H. (1998). Collective dynamics of 'small-world' networks. *Nature, 393*(6684), 440–442. https://doi.org/10.1038/30918.

Yose, J., Kenna, R., MacCarron, P., Platini, T., & Tonra, J. (2017). A networks-science investigation into the epic poems of Ossian. *Advances in Complex Systems, 19*(04n05), 1650008. https://doi.org/10.1142/s0219525916500089.

Did Homer write both the *Iliad* and the *Odyssey*?

21

Eric L. Altschuler[1,2]
[1]Essen Health Care, Bronx, NY, United States, [2]Kessler Institute for Rehabilitation, West Orange, NJ, United States

One question in studying and learning about Homer is to ask whether the same author wrote both the *Iliad* and the *Odyssey*. While this may seem obvious that there is a single author for both epics, nevertheless, it is worth examining the evidence.

Qualitatively to zeroth order, there is a single author as there are not two or more different claimants to have written the epics. Now, as we do not know anything about Homer one could argue that there is not even a single claimant. That having been said, there are certainly not *two* claimants and there was one important opportunity for a second author to step forward—see below.

To first order even in translation the language, style, characters, and subject matter—granted the *Iliad* is about a war siege and the *Odyssey* is a travel adventure story—are very similar in the two epics. To second order when scholars and linguists look at the epics they seem quite similar. The *Odyssey* seems most surely to have been written after the *Iliad* as the *Odyssey* has full knowledge of what happened in the *Iliad*—see below for details—but the *Iliad* belies no knowledge of the events in the *Odyssey*. These days one can wonder if to second- or third-order AI can be used to study the question of a single author. However, the lack of any other text to act as an "out group" would seem to preclude this though perhaps comparison of the epics at the level of individual books may yield useful information.

Is there other evidence to weigh in on one side or the other of the single-author question? One thing we certainly can do is to compare injuries in the *Iliad* and *Odyssey* to look for similarities, or lack thereof. I agree with Chistos Koutserimpas (see Chapter 7) that the pattern and description of injuries in the *Iliad* compared with the *Odyssey* is the same. (Left as an exercise to the reader is to compare the injuries in the *Odyssey* listed in Table 7.1 to those in the *Iliad* (Chapters 9–13).

In terms of comparing the two epics as works of literature Richard Lattimore in just five pages and six footnotes in the Introduction to his translation of the *Odyssey* (Homer, 2007) does an outstanding job, and I completely agree with him, in showing that either the *Odyssey* (along with the *Iliad*) was either also written by Homer or by someone who had studied the *Iliad* in all its details, large scale features and style. How does he come to this conclusion?

An outline of the *Odyssey* is given in Box 21.1. Note in particular the first four books—the so-called Telemachy which covers the coming of age and education of Odysseus' son Telemachus while Odysseus is still away on his wanderings after the fall of Troy. Many commentators have suggested that these books do not fit in well with the rest of the *Odysse*y and were not written by Homer. Conversely, Lattimore

Medicine in Homer. DOI: https://doi.org/10.1016/B978-0-443-32978-4.00008-0

Box 21.1 Outline of the *Odyssey*

An outline of the basic structure of the *Odyssey* is given in Table 21.1. A couple of complementarities of the epics are also worth noting: (1) the *Iliad* is a war epic while the *Odyssey* is a travelogue and love story. (2) The *Iliad* is a story about armies, while the *Odyssey* is a naval story and ultimately—with Odysseus' "landing on Ithaca" with Telemachus and storming the palace there to vanquish Penelope's suitors—also a tale of marines. Author's own work.

Table 21.1 Structure of the *Odyssey*.

Books of the *Odyssey*	Summary of action	Comments
1–4	Education and coming of age of Odysseus' son Telemachus	The Telemachy
5–12	Wanderings of Odysseus	
13–24	Odysseus' return to Ithaca and eventual reunion with Telemachus and his wife Penelope	

Source: Author's own work.

finds the Telemachy a most organic bridge of the *Iliad* to the *Odyssey* stylistically and plotwise. For example, favorite characters such as Helen, Menalaus, and Nestor return and reassume their prior personas: Helen as self-centered, Menalaus always courteous with a strong moral sense and garrulous Nestor. The Telemachy sets up later action in the *Odyssey* and of course develops the character of Telemachus who is mentioned in the *Iliad* as Odysseus' son (Table 21.1).

Next Lattimore delineates with examples classes of similarities between the two epics: Lattimore starts with similar phrases used in the *Odyssey* and *Iliad* in similar ways Table 21.2. Now, hypothetically, the similarities could be mere coincidences, the only way to state a certain point given metric constraints or an homage by some putative other author of the *Odyssey* to Homer (the author of the *Iliad*).

However, certainly not coincidences, products of metrical requirements or an homage by anyone except a totally devoted student and protégé of Homer (the ultimate "workshop" member) are points made **in the** *Odyssey* about the **time period covered by the** *Iliad*. As Lattimore notes the *Iliad* belies no knowledge of the *Odyssey*,[1] while when the *Odyssey* does refer to action in and the time period of the *Iliad* material is not repeated. That is, the author of the *Odyssey* knew the *Iliad* most intimately. When the *Odyssey* does speak of events during the time of the *Iliad* (Table 21.3), it is to make certain additional points of emphasis (some major like the

[1] Though, as I point out in Chapter 4, Odysseus' speech in the Thersites episode of Book 2 of the *Iliad* strongly hints at the main theme of the *Odyssey*.

Table 21.2 Similar phrases used in similar ways in the *Iliad* and the *Odyssey*.

Iliad ref.	*Iliad* text	*Odyssey* ref.	*Odyssey* text	Comments
Book 2, 434	ʻΑτρεΐδη κύδιστε ἄναξ ἀνδρῶν Ἀγάμεμνον, "Most glorious son of Atreus, Agamemnon, king of men"	Book 11, 397	ʻΑτρεΐδη κύδιστε, ἄναξ ἀνδρῶν Ἀγάμεμνον, "Most glorious son of Atreus, Agamemnon, king of men"	Agamemnon hailed by Odysseus the same way in the *Iliad* and *Odyssey*
Book 2, 50–52	αὐτὰρ ὃ κηρύκεσσι λιγυφθόγγοισι κέλευσε κηρύσσειν ἀγορὴν δὲ κάρη κομόωντας Ἀχαιούς: οἳ μὲν ἐκήρυσσον, τοὶ δ᾽ ἠγείροντο μάλ᾽ ὦκα: but Agamemnon bade the clear-voiced heralds summon to the place of gathering the long-haired Achaeans. And they made summons, and the men gathered full quickly.	Book 2, 6–8	αἶψα δὲ κηρύκεσσι λιγυφθόγγοισι κέλευσε κηρύσσειν ἀγορήνδε κάρη κομόωντας Ἀχαιούς. οἱ μὲν ἐκήρυσσον, τοὶ δ᾽ ἠγείροντο μάλ᾽ ὦκα. Straightway he bade the clear-voiced heralds to summon to the assembly the long-haired Achaeans. And the heralds made the summons, and the Achaeans assembled full quickly	Summons to assembly
Book 1, 73	ὅ σφιν ἐὺ φρονέων ἀγορήσατο καὶ μετέειπεν: He with good intent addressed the gathering, and spoke among them	Book 2, 160	ὅ σφιν ἐὺ φρονέων ἀγορήσατο καὶ μετέειπε: He with good intent addressed the gathering, and spoke among them	Introduction of a speaker
Book 15, 479–481	αὐτὰρ ὅ γ᾽ ἀμφ᾽ ὤμοισι σάκος θέτο τετραθέλυμνον, κρατὶ δ᾽ ἐπ᾽ ἰφθίμῳ κυνέην εὔτυκτον ἔθηκεν ἵππουριν, δεινὸν δὲ λόφος καθύπερθεν ἔνευεν: but about his shoulders put a fourfold shield, and upon his mighty head set a well-wrought helmet with horse-hair crest; and terribly did the plume nod from above	Book 22, 122–124	αὐτὸς δ᾽ ἀμφ᾽ ὤμοισι σάκος θέτο τετραθέλυμνον, κρατὶ δ᾽ ἐπ᾽ ἰφθίμῳ κυνέην εὔτυκτον ἔθηκεν, ἵππουριν, δεινὸν δὲ λόφος καθύπερθεν ἔνευεν: For himself, he put about his shoulders a fourfold shield, and set on his mighty head a well-wrought helmet with horse-hair plume, and terribly did the plume wave above him	Battle formula repeated

Source: Greek and translations from (Homer (1919, 1924).

Table 21.3 Additions, clarifications, corrections, notes, extra points, and buffeting of main points in the *Iliad* with new material from the *Odyssey*.

Iliad ref.	*Iliad* text	*Odyssey* ref.	*Odyssey* text	Comments
Book 4, 271–289		Book 8, 499–520 and Book 11, 523–537		Trojan horse and the final battle for Troy
Book 10, 254–578		Book 4, 240–264		Odysseus' spying expedition (solo, in the *Odyssey*), NOT with Diomedes (*Iliad*)
Book 23, 707–737		Book 4, 341–344		Wrestling match with Philomeleides of Lesbos (*Odyssey*), NOT with Aias (*Iliad*)
Book 11, 401–488		Book 5, 308–310		Odysseus' fight in the defense of the body of Achilles (*Odyssey*), rather than his fight with Trojans after other Achians had fled (*Iliad*)
Book 1, 1–305		Book 8, 75–82		Quarrel of Odysseus and Achilles (*Odyssey*) rather than Agamemnon and Achilles (*Iliad*)
Book 17- Book 18, line 225		Book 24, 35–94		not covered in *Iliad*: Death and burial of Achilles, rather than the death and burial of Patroculus
		Book 11, 541–564		Not in the *Iliad*: Quarrel between Odysseus and Aias over armor of Achilles
		Book 4, 187–88, 199–202		Not covered in *Iliad*: Death of Antilochos
		Book 8, 209; Book 11, 505–537		*after *Iliad** like a post-great war/pandemic type exploits of heroes who reached Troy after the action of the *Iliad* was over Philoktetes (Book 8); Neoptolemos, Eurypylos, and Memnon (Book 11)
		Book 24, 114–119		Not covered in the *Iliad*: The recruiting of the heroes

Source: Author's own work.

story of the Trojan horse!), clarifications, slight corrections or to add additional information that did not fit into the prior narrative.

While Table 21.3 illustrates narrative unity of the two Epics, Tables 21.4 and 21.5 speak to a single author crafting both works. In Table 21.4, we see examples where material from the *Iliad* is reused, in a different context, in the *Odyssey*. In Table 21.5, Homer goes a bit further repurposing material from *Iliad* for use in the *Odyssey*. In a not completely dissimilar way composers, for example, JS Bach (1685–1750) would often reuse (with small or large) revisions music written for one occasion or context for another.[2] See Figs. 21.1 and 21.2 for Bach's arrangement of an instrumental piece, the third movement of the first Brandenburg Concerto ((BWV 1046/3, 1723) (Fig. 21.1) (BWV 1046 - Brandenburg Concerto No. 1 (Scrolling), 2013) as the opening movement of the secular vocal cantata Auf, schmetternde Töne der muntern Trompeten (*Arise, blaring tones of high-spirited trumpets*) (BWV 207a/1, 1735; Fig. 21.2; Bach - Cantata Auf, schmetternde Töne BWV 207a - Van Veldhoven I Netherlands Bach Society, 2020).

Bach would often reuse sacred music in a different context (Altschuler, 2024), for example, the last movement of the Mass in F Major (BWV 235/6) utilizes material from the opening chorus of the cantata *Darzu ist erschienen der Sohn Gottes* (*For this the Son of God appeared*) (BWV 40/1), and he was not above repurposing secular music for sacred duty as in his adaptation of the opening movement of *Preise dein Glücke, gesegnetes Sachsen* (*Praise your good fortune, blessed Saxony*) a cantata written for the anniversary of the election of August III, Elector of Saxony, as King of Poland in 1734 (BWV 215/1) for the Osanna of the B Minor Mass (BWV 232/19, 21 (repeat), 1749).

Any discussion of authorship naturally brings up the question of how the Epics were written. One way to figure this out would be to find Homer's notes or working drafts. Unfortunately these do not exist. Indeed, while Bernard Knox in his Introduction to Robert Fagles' outstanding translation of the *Iliad* (Homer, 1998) does a good job of tracing printed copies of the *Iliad* back to Florence in 1488, and the manuscripts that must have been used to produce this volume, the trail eventually goes cold. There is a dispute about whether there were papyrus scrolls containing books of the *Iliad*. However, I think there must have been, if only as they were the sources that allowed intensive study of the *Iliad* and *Odyssey* by ancient scholars/grammarians (see Chapter 5).

Another school of thought, made famous by Harvard classicist Milman Parry (1902–1935) (who met an unfortunate early death in a hotel room in Los Angeles) and his student Albert Lord, is that the Epics are the product of a long tradition of oral narrative storytellers (Lord, 1960). The theory is based on their study of the works Yugoslavian itinerant bards and Homer's repeated use of stock adjectives to describe characters, for example, "Zeus as Lord of Gods and men," "Nestor the horseman of Gerenia," and "Diomedes good at the war cry."

To begin with not only is there no evidence for a tradition of Greek epic oral poetry preceding Homer, there isn't one following Homer either. The two Epics stand alone

[2] so-called "parodies"

Table 21.4 Combat language in *Iliad* used in a noncombat context in *Odyssey*.

Iliad ref.	*Iliad* text	*Odyssey* ref.	*Odyssey* text	Comments
Book 7, lines 268–269	δεύτερος αὖτ' Αἴας **πολὺ μείζονα λᾶαν ἀείρας** **ἧκ' ἐπιδινήσας, ἐπέρεισε δὲ ἶν' ἀπέλεθρον,** Then Aias in turn **lifted on high a far greater stone, and swung and hurled it, putting into the cast measureless strength;**	Book 9, lines 537–538	αὐτὰρ ὅ γ' ἐξαῦτις **πολὺ μείζονα λᾶαν ἀείρας** **ἧκ' ἐπιδινήσας, ἐπέρεισε δὲ ἶν' ἀπέλεθρον,** But the Cyclops **lifted on high again a far greater stone, and swung and hurled it, putting into the throw measureless strength**.	
Book 12, lines 384–386 (see Case 17, Table 13.1)	θλάσσε δὲ τετράφαλον κυνέην, **σὺν δ' ὀστέ' ἄραξε** **πάντ' ἄμυδις κεφαλῆς: ὃ δ' ἄρ' ἀρνευτῆρι ἐοικὼς** **κάππεσ' ἀφ'** ὑψηλοῦ πύργου, **λίπε δ' ὀστέα θυμός**. Aias lifted it on high and hurled it, and he shattered the four-horned helmet, **and crushed together [385] all the bones of the head** of Epicles; **and he fell like a diver** from the high wall, **and his spirit left his bones**.	Book 12, lines 412–414 (see Case 10, Table 8.1	πλῆξε κυβερνήτεω κεφαλήν, **σὺν δ' ὀστέ' ἄραξε** **πάντ' ἄμυδις κεφαλῆς: ὁ δ' ἄρ' ἀρνευτῆρι ἐοικὼς** **κάππεσ' ἀπ'** ἰκριόφιν, **λίπε δ' ὀστέα θυμὸς** ἀγήνωρ. On the stern of the ship the mast struck the head of the pilot and **crushed all the bones of his skull together, and like a diver he fell** from the deck **and his proud spirit left his bones**	
Book 15, 648	**σμερδαλέον κονάβησε** περὶ κροτάφοισι πεσόντος. **and terribly rang [from the preceding line: his helmet]** around his temples as he fell. (translation by F. Horn, used with permission)	Book 17, 542	**σμερδαλέον κονάβησε:** γέλασσε δὲ Πηνελόπεια, **and all the room round about echoed wondrously**. And Penelope laughed	

Source: Author's own work.

Table 21.5 Homer's parodies.

Iliad ref.	*Iliad* text	*Odyssey* ref.	*Odyssey* text	Comments
Book 12, 299–301	βῆ ῥ' ἴμεν ὥς τε λέων ὀρεσίτροφος, ὅς τ' ἐπιδευὴς [300] δηρὸν ἔῃ κρειῶν, **κέλεται δέ ἑ** θυμὸς ἀγήνωρ **μήλων πειρήσοντα καὶ ἐς πυκινὸν δόμον ἐλθεῖν:** so went his way like a mountain-nurtured lion [300] that hath long lacked meat, **and his** proud spirit **biddeth him go even into the close-built fold to make an attack upon the flocks**.	Book 6, 133–134	ἠὲ μετ' ἀγροτέρας ἐλάφους: **κέλεται δέ ἑ** γαστὴρ **μήλων πειρήσοντα καὶ ἐς πυκινὸν δόμον ἐλθεῖν:** into the midst of the kine he goes, or of the sheep, or on the track of the wild deer, **and his** belly **bids him go even into the close-built fold, to make an attack upon the flocks**.	
Book 6, 244–46	κοιμῶντο Πριάμοιο παρὰ μνηστῇς ἀλόχοισι, κουράων δ' ἑτέρωθεν ἐναντίοι ἔνδοθεν αὐλῆς δώδεκ' ἔσαν τέγεοι θάλαμοι ξεστοῖο λίθοιο therein the sons of Priam were wont to sleep beside their wedded wives; and for his daughters over against them on the opposite side within the court were twelve roofed chambers of polished stone,	Book 14, 13–15	ἔντοσθεν δ' αὐλῆς συφεοὺς δυοκαίδεκα ποίει πλησίον ἀλλήλων, εὐνὰς συσίν: ἐν δὲ ἑκάστῳ πεντήκοντα σύες χαμαιευνάδες ἐρχατόωντο, set close together, which he had made by splitting an oak to the black core; and within the court he had made twelve sties close by one another, as beds for the swine, and in each one [15] were penned fifty wallowing swine, females for breeding;	
Book 6, 493	**πᾶσι, μάλιστα δ' ἐμοί**, τοὶ Ἰλίῳ ἐγγεγάασιν.	Book 1, 359–360; Book 21,	μῦθος δ' ἄνδρεσσι μελήσει	

(*Continued*)

Table 21.5 (Continued)

Iliad ref.	*Iliad* text	*Odyssey* ref.	*Odyssey* text	Comments
	"… **for all, but most of all for me**, of them that dwell in Ilios."	352–353 and Book 11, 352–353	**πᾶσι, μάλιστα δ' ἐμοί:** τοῦ γὰρ κράτος ἔστ' ἐνὶ οἴκῳ. "… but speech shall be for men, for all, **but most of all for me**; since mine is the authority in the house." τόξον δ' ἄνδρεσσι μελήσει πᾶσι, μάλιστα δ' ἐμοί: τοῦ γὰρ κράτος ἔστ' ἐνὶ οἴκῳ. "… The bow shall be for men, for all, but most of all for me; since mine is the authority in the house." πομπὴ δ' ἄνδρεσσι μελήσει **πᾶσι, μάλιστα δ' ἐμοί**: τοῦ γὰρ κράτος ἔστ' ἐνὶ δήμῳ. "… His sending shall rest with the men, with all, **but most of all with me**; for mine is the control in the land."	

Source: Greek and translations from Homer (1919, 1924). These works licenced under a Creative Commons Attribution-ShareAlike 3.0 United States License.

(A)

Figure 21.1 BWV_1046_3_16_29. Bach: Brandenburg Concerto No. 1, Allegro (BWV 1046/3), measures 16–29.

and seem to come out of nowhere. Next, the Yugoslavian tales are not of the scope or quality of the *Iliad* or *Odyssey*. Finally, as cataloged by Lattimore in his excellent Introduction to his translation of the *Iliad* (Homer, 1961), Homer deployed a variety

(B)

Figure 21.1 *(continued)*

of different epithets for a given character based on the place in the meter/line the character's name was used. See Box 21.2. Homer needed the epithets to "get to the end of the line." Even today professors, dignitaries, and others are introduced consistently with various titles, honors, and other descriptors as the occasion requires and permits. Indeed, examples such as those in Tables 21.4 and 21.5 are more consistent with the eighteenth-century North German compositional planning than oral poetry.

Further to this notion that the *Iliad* and *Odyssey* are the deliberate compositions of one man are the findings and conclusions that can be drawn from a careful and thought-provoking paper by Saunders (2006). Saunders studied the strike verbs Homer used in association with various weapons. He found that Homer used the verb βάλλω ("throw," "cast") almost exclusively as the strike verb when the weapon was a missile such as a javelin or thrusting spear, but conversely used the verb ἐλαύνω ("beat," "drive") when a downward blow with a sword or axe is described (or implied) by the anatomy of a strike is a duel. Saunders goes on to show that ἐλαύνω is used in other contexts to indicate a downward movement, for example, a downward whipping of the reigns by a charioteer. So Homer is carefully matching and choosing his words to fit the meaning he intends in a given situation, rather than just reaching for stock "prepared" phrases.

Similarly, in a very recent paper, Folit-Weinberg (2022) has examined Homer's use of the words keleuthos and ὁδός (road, path) and κέλευθος (way, path). alkdsjf;asdf Folit-Weinberg notes that κέλευθος is used 86 times in the singular in the *Iliad* and the Odyssey and only once in the plural; lakdjf;ladsfa He show that Homer uses κέλευθος (keleuthos) to represent a journey or path, for example a literal

(A)

Figure 21.2 BWV207a_1_16_29 Bach: Cantata Auf, schmetternde Töne der muntern Trompeten, opening chorus (BWV 207a/1), measures 16–29.

(B)

Figure 21.2 *(continued)*

(C)

Figure 21.2 (*continued*)

Box 21.2 Homer's use and reuse of epithets

Lattimore (Homer, 1961) notes that Parry's argument for oral composition of epics rests primarily on Homer's use and reuse of stock epithets (Parry, 1928, 1930, 1932). Lattimore disagrees with Parry's contention that Homer was not able to make up all the epithets himself and had to rely on an oral singing tradition for these. I agree with Lattimore on this point. Lattimore questions whether Homer was able to both make up all of the epithets and invent the meter dactylic hexameter. I would suggest that Homer did not necessarily have to invent dactylic hexameter, but rather he could have appreciated the full potential of a meter in use for oral tales. Lattimore explains a given warriors/heroes will be referred to by varying epithets not based necessarily of the specific context of action in the Epic, but rather on Homer's metrical need for the line. For example, if Homer needs the last two feet to end a line–typically done (see Box 5.2) with a dactyl (– ◡◡, long, short, short) and then a spondee or trochee (– –, long, long or – ◡long, short) here are some epithets for heroes he uses: *Iliad*, Book 1, line 145ἢ Αἴας ἢ Ἰδομενεὺς ἢ **δῖος Ὀδυσσεὺς**either Aias, or Idomeneus, or **brilliant Odysseus**, *Iliad*, Book 6, line 466 ὣς εἰπὼν οὗ παιδὸς ὀρέξατο **φαίδιμος Ἕκτωρ**: So saying, **glorious Hector** stretched out his arms to his boy *Iliad*, Book 2, line 197 τιμὴ δ' ἐκ Διός ἐστι, φιλεῖ δέ ἑ **μητίετα Ζεύς**.for their honor is from Zeus, and **Zeus, god of counsel**, loveth them. or *Iliad*, Book 9, line 419 Ἰλίου αἰπεινῆς: μάλα γάρ ἑθεν **εὐρύοπα Ζεὺς** on the sheer city of Ilion since **Zeus of the wide brows** has strongly (Scansion of the lies is left as an exercise for the reader. See Box 5.2 for the basics of Homer's meter.) When Homer needs the last two-and-a-half-feet of a line (◡◡ – ◡◡ – –, short, short; long, short, short; long, long) he uses different epithets: *Iliad*, Book 1, line 311 εἷσεν ἄγων: ἐν δ' ἀρχὸς ἔβη **πολύμητις Ὀδυσσεύς**.and **Odysseus of many wiles** went on board to take command. *Iliad*, Book 6, line 342 ὣς φάτο, τὸν δ' οὔ τι προσέφη **κορυθαίολος Ἕκτωρ**: So said he, and **Hector of the flashing helm** answered him not a word*Iliad*, Book 14, line 312 τὴν δ' ἀπαμειβόμενος προσέφη **νεφεληγερέτα Ζεύς**: Then in answer spake to her **Zeus, the cloud-gatherer**. Finally, when Homer needs to fill the last three and one third feet (◡ –◡◡ –◡◡ – –) he deploys yet a different epithet for Zeus *Iliad*, Book 1, line 544 τὴν δ' ἠμείβετ' ἔπειτα **πατὴρ ἀνδρῶν τε θεῶν τε**: In answer to her spoke **the father of men and gods**: So, as Lattimore notes, Zeus goes from being a counselor, to a cloud gathering and then to the father of men and gods not based on any change in personality or attributes of Zeus, or on the circumstances of the situations of these lines (left as an exercise to the reader to check this point), but rather based on Homer's metrical need. Indeed, furthermore, metrical need/expediency leads Homer to use the same epithet for different characters: *Iliad*, Book 5, line 855 δεύτερος αὖθ' ὡρμᾶτο **βοὴν ἀγαθὸς Διομήδης** Next **Diomedes, good at the war-cry**, drave. *Iliad* Book 15, line 656 ὣς ἔφατ', οὐδ' ἀπίθησε **βοὴν ἀγαθὸς Μενέλαος**, So spake he, and **Menelaus, good at the war-cry**, failed not to hearken. As Lattimore comments, it was not, almost surely, that Diomedes and Menelaus were particularly louder than other warriors, but rather that both their names scan short, short, long, long. And Homer uses the epithet "good at the war cry" commonly for both heroes. Now, these two points of a-using epithets for characters based on the metrical need of a given line and b-using the same stock epithet for multiple characters might be seen as very strong evidence in favor of the long oral tradition theory for composition of the epics. However, just as much, if not even more to me, use of metrically convenient stock epithets could support the theory of a single author for the Epics: One person composing such long and complex works needs to save some time/mental space by using stock epithets, given the confines of the meter. And Homer certianly does not always use stock epithets for heroes. For example,*Iliad*, Book 17, line 513 **Ἕκτωρ** Αἰνείας θ', οἳ Τρώων εἰσὶν ἄριστοι. **Hector** and Aeneas, the best men of the Trojans. Indeed, in the example below Homer in line 402 (*Iliad*, Book 17) does use the stock epithet δῖος (goodly/brilliant) to describe Achilles the same epithet he used to describe Odysseus in line 145 of Book 1 of the *Iliad* (see above). However, Zeus in line 400 is not described by any epithet. It seems that names and epithets are used deliberately by Homer as context and meter necessitate—deliberate composition of an individual rather than being the result of ongoing oral extemporizations of a tale: *Iliad*, Book 17, lines 400–402 τοῖον **Ζεὺς** ἐπὶ Πατρόκλῳ ἀνδρῶν τε καὶ ἵππων/ἤματι τῷ ἐτάνυσσε κακὸν πόνον: οὐδ' ἄρα πώ/τι/ ᾔδεε Πάτροκλον τεθνηότα **δῖος Ἀχιλλεύς**:Such evil toil of men and horses did **Zeus/** on that day strain taut over Patroclus. /Nor as yet did **goodly Achilles** know aught of Patroclus' death Author's own work.

one (typically by water or air) such as Odysseus' ship back to the Greek camp after delivering Criseis to her father the preist Crises (see Box 5.1)[3]:

Iliad, Book 1, line 483

ἣ δ' ἔθεεν κατὰ κῦμα διαπρήσσουσα **κέλευθον**.

and she [the ship] sped over the wave, accomplishing her **keleuthos**.

A more figurative path described as a κέλευθος (keleuthos) in Book 12 of the *Iliad* (lines 410–411) is one Trojan ally Sarpedon King of the Lycians see adslk;fj is seeking through the wall the Greeks have erected to protected their ships:

ἀργαλέον δέ μοί ἐστι καὶ ἰφθίμῳ περ ἐόντι
μούνῳ ῥηξαμένῳ θέσθαι παρὰ νηυσὶ **κέλευθον**:
Hard is it for me, how mighty so ever I be,
alone to breach the wall, and make a **keleuthos** to the ships.

Conversely ὁδός is used to denote a physical road, such as on Odysseus' men took in Book 10 of the *Odyssey* (lines 103–104):

οἱ δ' ἴσαν ἐκβάντες λείην **ὁδόν**, ᾗ περ ἄμαξαι
ἄστυδ' ἀφ' ὑψηλῶν ὀρέων καταγίνεον ὕλην,
Now when they had gone ashore, they went along a smooth **hodos** by which wagons were wont to bring wood down to the city from the high mountains.

or more figuratively as a future journey as one Telemachus Odysseus' son will take in the *Odyssey* Book 2 (see Box 21.1), line 285

σοὶ δ' **ὁδὸς** οὐκέτι δηρὸν ἀπέσσεται ἣν σὺ μενοινᾷς:

But for thyself, the **hodos** on which thy heart is set shall not be long delayed

[3] Greek and translations from Homer (1919, 1924). These works licenced under a Creative Commons Attribution-ShareAlike 3.0 United States License.

The syllables of κέλευθος (keleuthos) scan short long short (◡ – ◡) so this is a word which goes well at the end of a line of dactylic hexameter, the last two feet the overwhelming majority of the time are a dactyl (– ◡◡, long, short, short) followed by a trochee (– ◡, long, short) with the first syllable of κέλευθος (keleuthos) making up the end of the fifth foot of the line and the last two syllables making up the sixth foot. See both *Iliad*, Book 1, line 483 and *Iliad*, Book 12, line 411 above. Homer most always uses κέλευθος (keleuthos) at this place in the line. ὁδός (hodos) which scans short, short (◡◡) Homer moves around in the line: In *Odyssey*, Book 10, line 103 ὁδόν (the ACCUSATIVE or direct object form/case of ὁδός (also scanning short-short)) makes up the second half of the fourth foot of the line, while in line 285 from Book 2 of the *Odyssey* ὁδός comprises the second half of the first foot. (Scanning the full lines from above is left as an exercise for the reader.)

Thus when composing the *Iliad* and the *Odyssey* Homer carefully and deliberately chose and arranged his words. The *Iliad* and the *Odyssey* therefore seem not the end result from a school of oral storytellers, but rather the product of Homer's carefully and deliberately chosen words.[4] Perhaps at some point, AI/LLMs could be used to try to discern between the two possibilities.

The accuracy and specificity of anatomic descriptions of injuries in the epics are also consistent with a single author. Any slight deviation or difference of text produced over time by an oral tradition method of composition over time would have led to the degradation of these.

The economic question of who paid for/funded the composition of the Epics is most simply answered with a single, physician author during time not spent making money practicing medicine. Indeed, the Epics certainly are not written from the partisan perspective/to the advantage of any king or city state that could have been a funder.

Finally, the end of the *Odyssey* is very weak from a literary point of view. As Lattimore notes, The Hand is gone. Surely if there was a second author of the *Odyssey* or even a crank claimant this would be the time to write a better ending and claim one was Homer (or some other name). But no one ever did. There was just one writer who wrote both epics.

References

Altschuler, E. L. (2024). *The neuroscience of Bach's music*. Elsevier.

Bach - Cantata Auf, schmetternde Töne BWV 207a - Van Veldhoven I Netherlands Bach Society. (2020). https://www.bachvereniging.nl/en/bwv/bwv-207a/.

[4] Classicist J.B. Hainsworth came to a similar conclusion (Hainsworth, J.B. The Flexibility of the Homeric Formula. (Oxford, 1968)). Further supporting the idea that in a calculated manner Homer deliberately deployed lexical devices to create flexibility within the structure of the meter, Hainsworth ("The Epic Dialect" in A Commentary on Homer's Odyssey, Volume I Alfred Heubeck, Stephanie West and J.B. Hainsworth Clarendon Press 1988, pp 24–32) notes the work of K. Witte ("Zur homerischen Sprache," Glotta i (1909) 132–145, ii (1910) 8–22, iii (1911/1912) 105–56) who pointed out that Homer uses an archaic form of the genitive singular -οιο—which scans as two syllables (–◡)—alongside the Ionic form of the day -ου scanning as only one syllable (–), but does not use the archaic form of the accusative singular -ονς which scan as one syllable (–) like the standard Ionic form -ους (–).

BWV 1046 - Brandenburg Concerto No. 1 (Scrolling). (2013). https://www.youtube.com/watch?v=7TlG6LMQR9M.

Folit-Weinberg, B. (2022). The language of roads and travel in Homer: Hodos and keleuthos. *The Classical Quarterly, 71*(1), 1–24.

Homer. (1919). *The Odyssey*. Harvard University Press.

Homer. (1924). *The Iliad.* Harvard University Press.

Homer (1961). *The Iliad of Homer*. University of Chicago Press.

Homer (1998). In B. Knox (Ed.). *The Iliad.* Penguin Classics.

Homer. (2007). The Odyssey of Homer.

Lord, A. B. (1960). *The Singer of Tales*. Harvard University Press.

Parry, M. (1928). *L'Épithète traditionelle dans Homère: Essai sur un problème de style homérique*. PhD thesis. Sorbonne University, Paris.

Parry, M. (1930). Studies in the epic technique of oral verse-making. I. Homer and Homeric style. *Harvard Studies in Classical Philology, 41*, 73–147.

Parry, M. (1932). Studies in the epic technique of oral verse-making. II. The Homeric language as the language of oral poetry. *Harvard Studies in Classical Philology, 43*, 1–50.

Saunders, K. B. (2006). Sword-fighting in the Iliad: A note on ΕΛΑγΝΩ. *The Classical Quarterly, 56*(1), 279–784.

Posttraumatic stress disorder in Homer: a female veteran's perspective

22

Simone Swartz
Author and Service Dog Strong Founder, Portland, ME, United States

From a woman's perspective. From a mother's perspective. From a female soldier's perspective. From a wife of a combat soldier.

I have served in the Army. I have been married to someone who has seen combat and suffered PTSD (posttraumatic stress disorder) from it. I have suffered from PTSD from a sexual assault that happened to me in the military. I have dedicated my life to further advancing PTSD awareness and encouraging all affected populations with PTSD from combat, SA (sexual assault), gun violence or otherwise that recovery is possible and PTSD does not have to be a life sentence. I founded a nonprofit that gives service dogs to SA-PTSD survivors. Now I speak to the world about my personal experience overcoming PTSD and help others recover and do the same.

The two main themes of Homer's *Iliad* and *Odyssey* that I want to speak on are gender roles and PTSD. These themes were very apparent and overwhelming to me. The severity of violence in general made *The Iliad and Odyssey* very hard for me to get through. My expertise in PTSD made this book very appropriate for me to read and comment on. Let me share with you my thoughts and encourage anyone dealing with intrusive thoughts from traumatic events to seek professional help with no shame. Know that PTSD has been a problem for soldiers since Homer's time. No matter how well the literary genius can package war, such as Homer's *Iliad and Odyssey*, or how many awards can be won for a war movie of our time, the lasting damage of violence is real. Trauma is real to the body and the mind. PTSD, although not named in Homer's time, was still felt and noticed as troublesome behavior to some soldiers who returned home from war.

To read Homer and see the many pages that idealize war and violence, and to bring to light the turmoil losing comrades and loved ones can cause when lost to violence, it is apparent that treatment for the effects of violence is still in the developmental stage. We can see that medicine has come a long way in its treatment, yet it still has a long way to go until we can properly treat combat veterans or anyone with PTSD properly.

Homer raised more questions for me about humanity than a look back in time to the days in which he lived. In this chapter, I will be proposing many questions for you to ponder. From the start of the story, there is a stereotypical incident of a woman being captured and the strong man coming to rescue her. As a woman, I immediately was reminded of the oppression of women and how this story has not aged well. We have the strong men stereotype of coming to rescue the 'damsel in distress."

Medicine in Homer. DOI: https://doi.org/10.1016/B978-0-443-32978-4.00013-4

The typical story theme of a distressed woman needing to be rescued is hard to come to terms with from a woman's perspective. This story shows the difference between men and women.

I will admit that upon hearing that an entire war could be started over the desire to save one woman is a bit romantic sounding. But, is it really romantic to send thousands of men to die? Is violence met with violence the answer? Is war a civil way to handle disagreements? Are women who have beauty the only ones needing saving? I feel we are truly still barbarians if fighting is the answer to solve such issues.

Are males going to keep being judged on their fighting ability and its animalistic behavior for another 2800 years? Can we switch our engrained misogyny into viewing human beings and their value on how well they care for one another? Is this fighting mentality intrinsically evolving within us or is it a conditioned reaction? Are we going to keep viewing women as mere damsels in distress and valued only when they are young and fertile? Are we going to view men as heartless soldiers who have to prove their born gender by slaughter? Can we switch our ingrained misogyny into viewing human beings and their value on how well they care for one another?

Homer's books are similar to the age-old Grimm's Fairy Tales and Disney's eventual renderings of the same. Is a storyline of women needing to be rescued more appealing to men or women? Is this story line encouraging in the fact that women are valued enough to be rescued come hell or high water? Or only the most aesthetically beautiful woman is worthy of being rescued come hell or high water? Or does that worthiness only apply to the most aesthetically beautiful woman?

The story line of a big burly, strong man fighting till brutal death to protect a woman's blemish free beauty can be romantic in some sense and barbaric in another. Many questions arose for me as to whether women are worth fighting for? Or, do women truly have to be protected by men? The deep traditional gender roles are heavy in the *Iliad*. Even though some cultures hold women and men to different roles where women also would be on the battle-field, in the *Iliad* the men go to war and fight to save the woman.

We are still fighting the stereotypes of traditional gender roles today in 2024 (Brook, 2023).

I question how far gender stereotypes have really come in the past 2800 years, and how far will these stereotypes and women's rights progress going forward?

Women are sexualized in *Iliad* by placing a woman as "high value" by the evaluation of outer beauty alone. This was a social theme that women's beauty is equal to their worth and this ideal is still propagated today by the patriarchy. Is a woman's value and worth to be "saved" more significant if she is beautiful? Why would a human being's worth be directly in line with outer appearance which, to a high degree, cannot be controlled? Who is determining what true beauty is? Whole armies, multiple hundreds of thousands of men, have set foot to fight for a beautiful woman.

On the flip side, the male gender role is also amplified to the utmost. A man's ability to conquer, kill and have a beautiful woman by his side equals success and valor. Is this belief ingrained in us from birth, or has society imposed its norms from its past outdated history?

As a young girl, I would often play in the woods these reenactments of being the damsel in distress and imagining the typical "knight in shining amour" coming to save me and we would live "happily ever after." I did not desire or reenact myself fighting off intruders or having to defend myself. Was this just particular to time period in which I grew up in, was this a gender role I associated with, or did I grow up to seek safety in someone physically stronger than myself? Is it a typical gender role for women to want to be protected? Is it ingrained into our society from centuries of literature, such as the *Iliad*, perpetuating this scenario? Or is it just romanticism—a well-put-together storyline adding in excitement of battle?

As a young adult in 2006, I enlisted in the Army with exciting and scary thoughts of being sent overseas to fight the ongoing "war on terror" after 911. I was met with disappointment to find out that women were treated much differently than the male soldiers. Women were not allowed "on the front lines."

In 2024, it is my hope and belief that our society is infringing less and less upon assigned gender roles. We should be able to freely choose and live a life that aligns to ourselves with utmost freedom. If centuries of deep societal norms are preventing this from happening, we have a long way to go in changing such accepted views on men and women.

It can be argued that there were some nontypical gender-conforming persons in *Iliad*. But without the freedom to express their true individuality, we can see an internal conflict and uptake of anger.

Is it normal for men to fight? Is it normal for women to fight? Is Homer trying to deal with his trauma within these hundreds of pages by describing bloody violence? Or do we see a warrior's PTSD from constant and prolonged hypervigilance, increased adrenaline, and the fight or flight/freeze/fawn response from this life of perpetual danger? Are men more comfortable with violence or is it a learned behavior and fixation from centuries of human conditioning (possibly starting with the *Iliad*, as Homer is one of the first known manuscripts to describe the battle and classic relationships between a man and a woman)?

It seems Homer has speckled his Epic with heavy overtones of the classic view of women as being weaker and needing to be saved and judged most by our "fairness" and beauty.

Book 3 quotes "marvelously and divinely lovely' as a repeated theme and I began to get the feeling of a woman's worth being shrunken down to merely external aspects when only beauty is repeatedly brought up with no mention of education or wit. Though this book was written quite some time ago, we can, of course, expect the classic misogynistic view of a woman's physical worth to be upheld above all else. This misogynistic view of women is still being held today.

Women are typically smaller in build and body size. Smaller muscles or muscle strength compared to a male their age, typically. The woman who was taken captive was royalty and would probably not be involved in any type of hard manual labor. She was probably not taught self-defense or protection skills and could have been unable to fight off her abductor by herself (Female abduction statistics, 2024):

"Nearly 1 in 5 women globally have experienced some form of abduction or kidnapping during their lifetime.

- Women ages 18–24 are at the highest risk of abduction compared to other age groups.
- In the United States, an estimated 300,000 women are abducted annually.

- Over 70% of female abduction victims knew their abductor.
- Only 55% of abducted women are reported missing to law enforcement.
- The majority of female abduction cases involve sexual assault.
- Approximately 1 in 7 abductions of women involve restraining them in a confined space.
- Women from marginalized communities are at a higher risk of abduction.
- Around 25% of female abduction cases involve captors using social media to lure victims.
- Only 53% of abducted women are able to escape or are rescued.
- Women who are abducted are at a higher risk of developing post-traumatic stress disorder.
- Female abduction is more prevalent in war-torn countries and conflict zones.
- Women from low-income households are at a higher risk of abduction due to lack of resources for protection.
- Abductions of women have been linked to organized crime in various regions.
- Only 37% of abducted women receive proper mental health support after their ordeal."

Does it seem women are judged on beauty alone still today? Are women still targets for violence more than men due to physical stature? Are men judged on physical strength alone? Can men and women be judged off mental strength? Can a woman's emotional strength and intuition be just as vital to humanity as physical strength?

My question is, does this theme of women being physically less strong, does this mean our worth is less? Why have we not diverted from the classical misogynistic views that brute strength is the only valued skill? Are women going to be continually judged on beauty alone?

Book 2 of *Iliad* has strong PTSD references when Zeus' wife speaks with her husband about the war. The constant fighting and continual rehashing of violence and intricate details of the wounds are an eerie reminder of hearing a modern-day soldier relive their days deployed in the battlefield. PTSD is a recurrence of memories overtaking the present and causing disturbances. The overwhelming reiteration of violence leads me to speculate that Homer had PTSD and severely. The reliving and glorification of the intense trauma and violence shows me that war was on his mind continually. War should not be an easy thing to speak on and a thing to keep prolonging. From a combat PTSD standpoint, life revolves around death, war and fight for survival. This struggle is not only from war itself but the aftermath when the soldier comes home and is not actively fighting. The warrior with PTSD is still reliving the fight. Not only in their mind with worry, but in the body can manifest PTSD symptoms as well. The constant fawn, freeze, fight or flight mentality can make the body physically react as if the warrior was still there active in a battle.

I debate that Homer was a doctor or a warrior, during a war and he himself potentially suffered from PTSD. Would he feel that women are not needed for survival because women couldn't fight, but only for their beauty? Even the wife Juno, who was bright and genuinely concerned about her husband, I felt was passed off as annoying him or nagging, when she was merely using her intuition trying to care for her husband. Someone experiencing PTSD will push away tender emotions of their own and others and practice only the most primal of emotions, anger and desire. Others showing tender emotions to the person with PTSD can sometimes be overwhelming for persons with PTSD.

It has been roughly 2,800 years since the *Iliad* and *Odyssey* were written or transcribed from spoken poem onto paper. In this vast array of time, it seems human nature and the way women are treated and viewed has not changed much (Mattison & Boyles, 2023). Power and control have seemed to conquer ethics and equity for the past 3000 years.

We see a great PTSD response from the soldiers who shut off all human emotion and eliminate their enemy (a fellow human) during war. There is a quote that says 'The true test of a strong man is how well they handle adversity and how they respond to their opponents." Is the true test of a "man" how much they can kill without emotion? Is the ideal that "the bigger, the better" keep perpetuating? Does this patriarchal theme keep going because it only benefits the persons in power? Being a woman in America today, reading Homer was insightful on how we got to this point in today's society and showed me how far we still have to go if we want social change and our overall views on war and women to evolve into a more peaceful and less oppressed perspective than 400 years ago.

Reading Homer as a woman in America, I gained an insight on how we got to this point in our society. American women were only allowed to open a bank account in their own name in 1974. Women have been forced to be taken care of financially and their safety be taken care of by other men since Homer, so it is very apparent the oppression of women has been ongoing for too long (Adam & Aldrich, 2023).

Women in the American military were allowed by the oppressor to battle "on the front lines" in combat in 2013 (Brook, 2023).

Reading the *Iliad* has reminded me how far we have to go in women's rights. Homer showed how humans have not changed much in 3000 years. Trauma still highly affects even the strongest of soldiers. PTSD can be treated better today. The need to eliminate war and the aggrandizement of violence needs to start to diminish. The world can become a more peaceful place with a focus on mental health.

I was continuing to search in the story for some glimpse of humanity. Some glimmer of equity or even a short humorous break. Just one line in the stories to lighten the violence, a lighthearted joke, would you please, Homer? Apparently not. I did however pick up on the intensity of drama and death. Such as someone stuck in depression would always view the world as gloomy and reiterate their inner turmoil and recurring nightmares. Homer is nowhere near Shakespeare in leaving the audience with a few moments of distraction from the drama of the main theme.

From my experience with PTSD and living with someone with untreated PTSD I clearly make the connection that the person who wrote these stories does indeed have PTSD. The relentless rehashing of violence in graphic detail does not stop. The absence of emotion or theatrical aside, humor or details of warm emotions leads me to believe the person who wrote this story was facing their own repeated unwanted memories of firsthand witnessing or being involved in these traumatic events of war. The story screams to me of someone who has directly witnessed such violence in up-close detail. The PTSD overtones of conflict between interpersonal relationships and family life attest to my theory Homer had been involved in battle or provided medical attention to wounded warriors on the battlefield himself.

The theory that Homer was a doctor is highly believable. It would make sense he was partially unprepared to witness such atrocities, as if any person would be prepared to watch someone else die. But when a person witnesses such violence, the images can keep replaying over and over again for years to come. The witnessing of such violence and on such a grand scale can easily be seen as the author tries to escape his memories and process through the trauma by way of telling his tale through poetry.

Often those with PTSD can lose interest in things that once made them happy. They will turn toward darker habits or interests as a way to desensitize themselves with the violence they have witnessed. Without some sense of physiological help and time to heal from grief, a person can take an overzealous approach to violence in their unknown attempt to mask their pain and convince themselves what they have been through was normal. Page after page of war. Page after page of PTSD war memories becoming so real to me, and I was not even there.

Coming from a perspective of being a wife of a combat soldier, I have heard the same type of tales of violence and emotionless banter of "bravery" to have killed a fellow human.

I have woken up beside a husband who cannot sleep at night because of the memories of violence. I have told him it is a sign of strength to show emotion than to withhold it and act like violence is normal or it didn't affect you. I have been on the bitter end of the soldier's misplaced anger. He held onto anger being scared to release it because "men don't talk about their feelings" toxic mentality. Yet, even though they are safe at home and no longer in battle, their body and mind can still keep them in a state of fear which is displayed in many ways but peace and love it is not.

Instead of this incredibly long story of rehashing war to prove a man is a man for killing more men than someone else, maybe it can be a reminder that violence and war is traumatic and unhealthy. Seeking therapeutic medical attention after being in war or witnessing any type of violence is essential for a person's wellbeing. They do not have to live the rest of their lives reliving the trauma and violence if handled properly and dealt with as soon as possible. PTSD can leave huge mental wounds if not cared for. Reading through Homer seems to be a huge PTSD combat wound in print. Homer can be noted for being on the cutting edge of acute trauma care for his era, however, I also would like to add his contribution can be the basis for the need for mental health care along with physical wound treatment.

The *Iliad* and *Odyssey*'s contribution to society would not be possible without war. I still believe in the idea of a peaceful, warless society someday soon. I pray the world will soon be able to switch from a power and control model to a freedom and respect model. A world where the global economy profits off of health and wellness instead of pain and war. I believe in this hope and dream more than ever since reading Homer.

I would love to have seen, 2800 years later, a new scenario surrounding PTSD and gender roles. I would love to now live in a world in which how we treat each other is the true reflection of strength. Where dying over a country's flag or having to be traumatized for the rest of your life from a battle to defend imaginary lines would be counter intuitive. A world where a man would not feel the need to assert himself with bigger weapons and more violence.

Untreated PTSD and mental health wounds can be more painful and detrimental than physical ones. I believe these stories were Homer's failed attempts to get out of his mind the painful, graphic memories of war and try to process the trauma. Violence infects us all. Untreated PTSD can infect us all. Mental health is important and should be "manly" to talk about.

My slogan is "PTSD is treatable." I stand behind that wholeheartedly. Enduring trauma is not easy and neither is living with the memories. Recovery is possible. I am happy to know that although society in general has not moved on from typical gender roles much since Homer's era, I do see headway in a positive direction surrounding mental health and PTSD research. Mental health awareness should be directed to all genders and should not carry shame for those asking for help (Adam, and Aldrich, 2023; Brook, 2023; Female abduction statistics, 2024; Mattison, and Boyles, 2023). The violence of war should be left on the battlefield and never carried into the home and society. War itself has never been beneficial to society. I hope that society and the medical community continue to examine and find solutions for PTSD. I hope the medical community examines the duality of physical and mental health support for soldiers and citizens who have PTSD. "PTSD is treatable!" and my last hope is for the next epic novel to be filled with peace and acceptance of all. Here's to our health and mental health!

References

Adam, J., & Aldrich, E. (2023). When could women open a bank account? https://www.forbes.com/advisor/banking/when-could-women-open-a-bank-account/.

Brook, T.V. (2023). Women banned from combat. https://www.usatoday.com/story/news/politics/2023/04/16/women-in-combat-military-progress/11548931002/.

Female abduction statistics. (2024). https://worldmetrics.org/female-abduction-statistics/.

Mattison, L. & Boyles, D. (2023). Homer's Odyssey | Origin & Background. https://study.com/academy/lesson/when-was-the-odyssey-written-homers-ancient-greece.html.

War and medicine in Homer: a combat veteran's perspective

23

Gregory T. Para
Sarasota Parrot Conservatory, Sarasota, FL, United States

October 12, 2024, I'm sitting in the living room with my father discussing many things as one would do with a parent who has a terminal cancer diagnosis. He had just informed the family several weeks ago and this is the first of many trips I will make from Florida to New York to visit and help take care of him before he passes.

My father is a Marine Corps veteran who served three combat tours in Vietnam. His father served in the Navy and saw action on a ship during WWII as was my mother's father who also served in the Navy and saw action on a ship during WWII.

My brothers and I all served in the Navy with two of us serving combat tours in Afghanistan attached to the Army. My two nephews joined with one serving in the Army and the other in the Navy. My son serves in the Air Force and was stationed in Jordan when the entry control point was attacked by terrorists killing multiple contractors. Remember the entry control point as it will come up in the *Iliad* as well.

As one can imagine the military often comes up as a subject of interest and common bond between all of us. Subjects range from politics, role of our military on the international stage, government waste and policy, funding, policy, leaders in the news, agreements and disagreements about current world events, and our role. Sharing some of the miserable living conditions we endured during deployments contrasted by the experiences we had during port calls or places of interest in our military travels.

Absent was the actual discussions of what we experienced, saw, or participated in during combat engagements. It seems almost universal, like an unspoken code that veterans rarely discuss their wartime experiences. Most veterans prefer not to relive those moments and society generally looks down on veterans who speak of their combat exploits often referring to them as someone exaggerating or bragging or trying to make themselves larger than life.

I've examined decades of conversations we had around the dinner table with Dad and the more serious conversations we had once my mother and our children were not around to hear. Again, we never spoke about what we had done or seen. But we found a wealth of common ground and shared experiences even though we all had served in different areas of the world in different conflicts with different political winds and larger-than-life actors on the world stage like General Eisenhower, General Patton, General MacArthur, Admiral Nimitz, Winston Churchill, Hitler, Joseph Stalin, Emperor Hirohito and Benito Mussolini, Nixon, General Westmoreland, Admiral Zumwalt, Ho Chi Minh, Ngo Diny Diem, Bush, and Obama similar to how Homer highlights Achilles, Agamemnon, Patroclus, Odysseus, Nestor, Priam, and Hector.

Medicine in Homer. DOI: https://doi.org/10.1016/B978-0-443-32978-4.00005-5

Common experiences we shared regardless of the conflict were the local peoples and our interactions with them, poor living conditions we endured and how misery loves company, learning the culture of the country where we were stationed, conversations with our interpreters, and learning about their social structure. We will delve more into patriarchal and matriarchal lineage and the commonality between the *Iliad* (how each side knew the families of the other) and how our interpreters knew this and compare it to present day how through spies inserted into populations that the United States and Allies know all the chain of command by name and family as well as other family members of our enemies and their roles in a particular conflict. We would discuss how things were made ranging in subject from our uniforms and bases to local made items and goods.

Introduction

For those who have gone to war there are many aspects of the *Iliad* that would suggest Homer witnessed these events and elegantly described in detail his experience. My study has shown many aspects of war never change and I will parallel events to help the reader see this perspective. Homer would not have had the information necessary unless he was there to articulate this poem as he did. In other words, there were no records, books, or previous wars to study in order to research for the *Iliad* yet he detailed every aspect of war from the preparation, forward deployment of troops, developed supply lines, ship stores, materials needed, operational knowledge of ships, how wages were paid, a myriad of military terms and finally battle scenes and injuries that would suggest Homer was present. Homer captures the human emotions from fear, to love, to anger, to longing for home, to family honor to rage, superstition, panic and morale, prayer, and despair.

As a former supply officer one aspect of the *Iliad* that captivated me was the listing of materials used during the war. Not only were there a myriad of materials listed but also what materials were used to create weapons of war. For instance, in my time in Afghanistan, there were two types of body armor worn. The difference is the construction and thickness of the Kevlar plates carried. Only someone intimate with this knowledge most likely having been around the troops would know this. Homer often cites specific construction of ships, chariots, and weaponry as well as armor suggesting he possessed more than a passing knowledge of these items. Here is one such example. (All translations from Homer (1924). This work is licensed under a Creative Commons Attribution-ShareAlike 3.0 United States License.):

> ***Book 7, lines 219–223***
> *So Aias drew near, bearing his shield that was like a city wall, [220] a shield of bronze with sevenfold bull's-hide, the which Tychius had wrought with toil, he that was far best of workers in hide, having his home in Hyle, who had made him his flashing shield of seven hides of sturdy bulls, and thereover had wrought an eighth layer of bronze.*

I would almost believe that Aias in conversation with others would have been heard bragging about his shield and where it was from and who made it. This would have been a common subject amongst the men. Regularly we would sit around and talk about the different boots we had found to bring with us or our weapon holsters that were better than our standard-issued gear or the different types of rifle slings we chose.

Homer exhibits more than a passing knowledge about ship construction and operation. Here is one such example:

> ***Book 1, lines 434–435***
> *When they had arrived within the deep harbour, they furled the sail, and stowed it in the black ship, and the mast they lowered by the forestays and brought it to the crutch with speed, and rowed her with oars to the place of anchorage. [435] Then they cast out the mooring-stones and made fast the stern cables, and themselves went forth upon the shore of the sea.*

Throughout the *Iliad* Homer again describes in detail the ship stores, supplies carried and their use, knowledge of chariot construction and repair as well as weapon design and construction.

The following are materials cited throughout the *Iliad* and I've broken them down into categories:

Materials for weapons

- Bronze (spear tips, spears, armor, pikes, chariot wheel spokes),
- Iron (battle ax, chains, chariot wheel axle),
- Horsehair (battle shields, clothing, helmet adornments),
- Wood (oak axle of chariots, ship repair, arrows, berm wall, used for funeral pyre, lances),
- Leather (rawhide shields, books, string),
- Gold (decorative items like chalices, battle gear adornment, bars for trade and ransom),
- Silver (decorative items like cups, plates, spear tips, chariot wheel hubs), and
- Rocks (for ammunition to throw during close encounter engagements, shore up base of wall).

Supplies carried on ships to maintain the troops

- Water,
- Livestock (sheep, swine, horses, cattle, goats, oxen, mules),
- Timber (ship repair, building rampart for protecting the ships on shore),
- Iron for axes and chains,
- Leather (oxhide and horsehair),
- Wine,
- Medicine (drugs, potions, herbs, and spices),
- Slaves,
- Wood (oak) for chariot axles,
- Bronze for chariot wheel spokes,
- Cloth, carpet, linen, cloaks, blankets,
- Oats and barley most likely for livestock feed,
- Chariots,
- Canvas (wrapping the chariots, clothing),
- Mercantile (carpet, cloth, linen, cloaks, blankets),
- Enamel (decorative pieces for furniture, helmets, shields),
- Protective gear (shields, helmets, footwear, clothing, body armor),

- Clothing (leather, canvas, wolf skin, weasel skin, cloth, linen), and
- Gold and silver (to pay ransoms and pay the troops).

Along the coast Homer points out that the Argives would stop in order to pillage towns for supplies and loot to pay the troops. This would suggest that Homer either heard about the exploits or was on the sea with the ships when they pulled into different ports. I would suggest he was on the ships as he cited specific numbers of ships and troop strength when he described the scene.

> ***Book 1, lines 124–126*** *"We know nothing of a hoard of wealth in common store, but whatever we took by pillage from the cities has been apportioned, [125] and it is not seemly to gather these things back from the army." This passage suggests they did not carry large amounts of wealth on the ships but would add to their wealth while sailing the coast.*

Homer gives us more than a glimpse of his detailed knowledge of ship construction and operations throughout this epic poem. Another layer of knowledge regarding weaponry used, weapon repair, chariot construction and repair is sprinkled throughout the poem as well.

The intimate knowledge of military terms, tactical operations, strategic direction and actual combat scenes suggests that the writer must have been there. To describe a scene where insubordination occurs, a salute is rendered, the weak of heart are placed in the center of a military formation so they cannot run from battle or a tactical flank is executed during battle would again suggest the writer would have witnessed these small nuances in an ongoing conflict. Taken individually there was no reason to add to the narrative as each had no impact on the story but collectively suggest Homer was a participant.

Here are military terms used in the Iliad

Troop formations: Front line, squad, rank, squadron, line of battle, columns and closed ranks.

Military terms: plan of action, muster, inspection, left flank, squadrons marshaled, call to arms, battle order, "dogs of war," infantry, "the press," tactician, strategist, salute, dishonorable discharge, battle order and "the troops."

During the *Iliad*, we find that Odysseus is considered the great tactician and Sarpedon a Trojan the master strategist of the Lycians.

An interesting passage is described by Homer in **Book 15, lines 347–350** "Whomsoever I shall mark holding aloof from the ships on the further side, on the very spot shall I devise his death, nor shall his [350] kinsmen and kinswomen give him his due meed of fire in death, but the dogs shall rend him in front of our city."

Here Homer is saying not to burn the dead combatant because the dogs will eat them. While I was in Afghanistan I witnessed the dogs at night coming to the walls of our camp and getting bones from the ground. These bones were from corpses long buried in the Soviet campaign against the Afghans.

The style in which the *Iliad* is written going back and forth between men and gods can cause you to miss many subtleties common in war. I ask the reader to consider some of the themes that are woven into the fabric of this epic poem.

There is a phrase widely attributed to US military chaplain William T. Cummings during World War II that says "There are no atheists in foxholes" Throughout the *Iliad*, we find the combatants continuously crying out to the gods asking for favors or asking why they have been cursed or forsaken. The fact that a character is calling to a higher power during tense moments is no different than our soldiers talking with their god. Our enemies do the same thing.

I found it fascinating that Homer was so knowledgeable of both the Argives and Trojans in regards to specific individuals and their lineage. In my experience I have found that a lot of stock is put into family power, social structure, social class and family name. This is no different in the *Iliad* and having this detailed, intimate knowledge would be expected.

I remember my Afghan interpreters and Afghan partners were always interested in hearing about my family and all the lineage. They were familiar with the Taliban leaders and their family lines which we find is no different in the *Iliad*. My interpreters would tell me the different patriarchal lines of different Afghan military members as well as share their long lines of lineage with their own families. However, it was forbidden to ask about their wives or daughters in their culture. But we do hear from several of the wives and daughters.

Homer points out that the Trojans use mercenaries from other lands in order to muster an army to ward off the Acheans. War does not change and we can read where this method was used in the Bible, Medieval times where different cities would congregate to fight the king and in modern times as well.

Another part of war that has never changed has been the use of trenches, walls and ramparts. Homer illustrates the use of this tactic when the Argives are in the open with no protection amongst their ships on the shore. The Argives dig a ditch according to Homer so that the charging Trojans would not be able to keep up momentum, thus making it easier to repel them. (See Box 23.1 for the description of the Greek Trench.) The ships definitely would have carried timber to this particular location specifically for ceremonial fires and burning corpses but also to build a protective wall around the landed ships and troops. We read that wooden poles were used, door blocks were installed so that the Argives could enter the compound but it could then be locked to thwart the enemy from entering. Remember earlier I told you to remember the term "entry control point." Well here we see the same tactic used in modern times with a different name but same purpose.

We also read that they used timber to brace the walls to make them more sturdy to again have the ability to stand up against the battering the enemy would use on the wall. Finally, we read that spears were planted into the trench to penetrate the flesh of charging men and horses prior to reaching the wall.

In my military studies at the Naval War College a frequent concept reared its head in every discussion regarding warfare in history. It is a term recognized as "fog of war" This concept permeates every plan any nation makes when playing war games and developing operations and strategy in any conflict. It is widely recognized

Box 23.1 The Greek Trench.

Book 7, lines 435–444

and they made about it a single barrow, rearing it from the plain for all alike; and thereby they built a wall and a lofty rampart, a defense for their ships and for themselves. And therein they made gates, close-fastening, that through them might be a way for the driving of chariots. And without they dug a deep ditch hard by, wide and great, and therein they planted stakes. Thus were they toiling, the long-haired Achaeans; and the gods, as they sat by the side of Zeus, the lord of the lightning, marveled at the great work of the brazen-coated Achaeans.

Book 9, lines 346–350

Nay, Odysseus, together with thee and the other princes let him take thought to ward from the ships consuming fire. Verily full much hath he wrought without mine aid; lo, he hath builded a wall and digged a ditch hard by. wide and great, and therein hath he planted stakes;

Book 12, lines 49–59

Hector went ever through the throng and besought his comrades, urging them to cross the trench. Howbeit his swift-footed horses dared not, but loudly they neighed, standing on the sheer brink, for the trench affrighted them, so wide was it, easy neither to o'erleap at a bound nor to drive across; for overhanging banks stood all about its circuit on this side and on that, and at the top it was set with sharp stakes that the sons of the Achaeans had planted, close together and great, a defense against foemen. Not lightly might a horse, tugging at the wheeled car, get within that circuit; but the footmen were eager, if they might achieve it.

that despite intricate planning the "fog of war" inserts an element that cannot be recognized ahead of time and causes leaders to think on the ground in the moment many times with unanticipated disastrous results.

Although the actual term "fog of war" was never attributed to anyone it is associated with a Prussian military theorist Carl von Clausewitz (1780–1831). In his work *On War* [*Vom Kriege*] (von Clausewitz, 1832) in Book 1, Chapter 3, titled "On Military Genius" Clausewitz writes about the "uncertainty of all information in war" and how commanders must navigate incomplete and misleading data. "Over time the idea of a metaphorical 'fog' to describe the chaos and unpredictability of war emerged from interpretations of his writings."

In the *Iliad*, we see the "friction of war" and the "fog of war" coming out in scenes that show just how many moving parts in a battle happen normally especially when the tide of battle turns and why. While not elaborating on the theme of friction it is worthy to note an article written titled "On war without the Fog" by Eugenia C Keisling (Kiesling, 2001). The following quote from the article illustrates my point

"Rejecting the friction-fog dichotomy allows a better understanding of what Clausewitz actually means by friction. Instead of a mental fog and physical friction, he guides us to see two different forms of friction. On one hand, friction encompasses the physical difficulties of moving and fighting armies. On the other, he links friction with intangible factors—fear, physical hardship and problems of information—that hamper the military commander."

Is it possible Homer may have been the first to describe the "fog of war" with his use of the word "mist" and the descriptions of fear, hardships, and poor information? It is worth further study. Homer uses the word "mist" mostly to describe the friction being experienced during multiple battle scenes. He uses it once to describe the death of a combatant stating the last thing he saw was the mist.

Book 5, 126–127 Here we find Athene speaking with Diomedes while he is praying for clarity before he goes against the Trojans. "And the mist moreover have I taken from thine eyes that afore was upon them, to the end that thou mayest well discern both god and man."

Book 5, 774–776 In the preceding passages we see the tide of battle turn as it often does in this epic poem and is attributed to the gods favoring one side or the other and creating confusion on the battlefield described as a mist. This mist would likely be the fog of war where one side gains an advantage attributed to the gods. But when they were come to the land of Troy and the two flowing rivers, where the Simoïs and Scamander join their streams, [775] there the goddess, white-armed Hera, stayed her horses, and loosed them from the car, and shed thick mist about them; and Simoïs made ambrosia to spring up for them to graze upon. Thus giving aid to the Argive warriors.

Book 15, 666–670 The Athenians find themselves clustered around their ships awaiting an attack by the Trojans. Here Nestor chides them on with a call to fight for their families and loved ones dead and alive. In other words to fight with honor especially when questioning their position. While anticipating an attack a thousand thoughts go through a person's mind and a great leader would recognize there may be confusion amongst the troops and find it necessary to get them all ready with one mindset. "So saying, he aroused the strength and spirit of every man, and from their eyes Athene thrust away the wondrous cloud of mist, and mightily did light come to them from either hand, [670] both from the side of the ships and from that of evil war." Probably the clearest example of referring to the "fog of war" can be found at the end of this line when we see Athene thrust away the mist from those positioned in defense of their ships but also giving the Athenians a unified morale when Homer says that "light come to them from either hand both from the side of the ships and from that of evil war" With so many moving parts in war there are many things that can go wrong including combatants questioning why they are there. Again, I find this line a compelling argument for possibly the first author to define the concept of the fog of war.

Book 16, 778–790 [778] "Now as long as the sun bestrode mid-heaven" it is clear that it is noon during this battle. Homer does not state that any other weather condition exists but goes on to state what I believe is the confusion sowed amongst combatants in the fog of war. This makes a strong case for the definition for the fog of war. "Now as long as the sun bestrode mid-heaven," It is clear that it is noon during this battle and if they can see the sun there is no literal mist. "so long the missiles of either side reached their mark, and the folk kept falling; but when he turned to the time for the unyoking of oxen, [780] then verily beyond their portion the Achaeans proved the better. Forth from out the range of darts they drew the warrior Cebriones from the battle-din of the Trojans, and stripped the armour from his shoulders; and Patroclus with fell intent leapt upon the Trojans. Thrice then leapt he upon them, the peer of swift Ares, [785] crying a terrible cry, and thrice he slew nine men. But when for the fourth time he rushed on, like a god, then for thee, Patroclus, did the end of life appear; for Phoebus met thee in the fierce conflict, an awful god. And Patroclus marked him not as he passed through the turmuoil, [790] for enfolded in thick mist did he meet him." Here, the friction and confusion in battle arises and gives way to the metaphorical "fog of war."

It is early November 2024 and I've completed my chapter while sitting bedside with my father. On November 19, 2024, my father passed away or as Homer would have said "the mist passed before his eyes." I was with him in the end and from the time he was diagnosed until death was barely 90 days.

Whether Homer participated in the campaign between the Greeks and Trojans it is my belief that Homer may have been the first to describe the eventual term "fog of war." I believe he was there. Metaphorically he suggests this chaos and unpredictability we know Clausewitz wrote about as a mist that comes about during different battles; of this, there should be no debate.

References

Homer (1924). *The Iliad.* Harvard University Press.

Kiesling, E. C. (2001). Fog on war without the fog. *Military Review; Fort Leavenworth. 81*, 84–87.

von Clausewitz, C. (1832). *On war.* Marie Sophia von Clausewitz.

Conclusion and future studies 24

Eric L. Altschuler[1,2]

[1]Essen Health Care, Bronx, NY, United States, [2]Kessler Institute for Rehabilitation, West Orange, NJ, United States

So we've come to the end. What have we learned about medicine in Homer and medicine and Homer? Let us work backwards fromq the **fourth and last section** of the book **Further Perspectives on Medicine in Homer**: In the previous chapter (Chapter 23) combat veteran Greg Para "red teamed" the question of whether the battle injury scenes and descriptions in the *Iliad* were, from his experience, written by someone who had seen combat as a physician or just heard about these at some port, bar or inn. He concluded that Homer did in fact see battle and illustrates this with vivid accounts from his own time in war. Furthermore, drawing on his background in logistics he discusses issues of supply necessary for the expeditionary force in the *Iliad* again concluding that the extent and level of detail of Homer's discussion of these are consistent with Homer having been deployed. Also, Greg shows that Homer may have been the first to describe in print the principle/concept of the "fog of war." In Chapter 22, veteran Simone Swartz looked at women, men, their relationships, and also posttraumatic stress disorder (PTSD) in Homer. She found that Homer has women and men assume the traditional gender roles. With respect to PTSD given the constant rehashing of violence in the *Iliad* Simone thinks Homer must have witnessed first hand, she feels that writing the *Iliad* was Homer's method of working through the trauma by way of reliving it all over again, in lieu of traditional talk therapy or other treatments for PTSD available today.

In Chapter 21, I concluded that Homer did indeed write *both* the *Iliad* and the *Odyssey*. In Chapter 20, Pádraig MacCarron, an Associate Professor in the Department of Mathematics and Statistics at the University of Limerick used a Facebook network approach to study the relationships among characters in the *Iliad* and the *Odyssey*. He found, among other results, that the social network of the *Iliad* behaves like a real social network. Another group (Miranda et al., 2018) using a somewhat different approach to the network analysis also found that the *Iliad*'s social network behaves like a real social network (whereas the *Odyssey*'s is more akin to a fictional network which becomes more like a real network when certain mythological characters are removed from the network). These results are consistent with the *Iliad* having been based on some real campaign in which the author participated. MacCarron also found that the social networks of the *Iliad* and the *Odyssey* are very highly connected. This is consistent with the two epics having been written by the same author (see Chapter 21 and the Introduction to Homer (2007)). In Chapter 19, using genetic methods Mark Pagel found ~c.762 BCE is the most likely

Medicine in Homer. DOI: https://doi.org/10.1016/B978-0-443-32978-4.00025-0

date for the composition of the Homeric epics with an early date of 1200 BCE strongly excluded.

The **third section** of the book **Neurology, Toxicology, Infectious Disease and Genetics in Homer** looked at nontrauma medicine in Homer. In Chapter 18, we saw that Homer knew his genealogies, and wanted to share them, like authors from other societies in antiquity and modern authors as well. In Chapter 17, Andreas Plaitakis examined plants, poisons, potions, and plagues in the Epics through the lens of modern science and medicine. In Chapter 16, I discussed what is a relatively complete description by Homer of antisocial personality disorder (ASPD) in the mythological hero Heracles. I also cataloged mythological stories in Homer already well-known by his audience. Chapter 15 covered exceptional neurologic and neurotrauma cases in Homer.

The **second section** of the book **Trauma in Homer**, the heart of the book, covered injuries in Homer by body area. In Chapter 14, Professor Fabian Horn, a Faculty Member in the Department of Greek and Latin Philology at Ludwig-Maximilians-Universität München discusses classicists' perspective on trauma in the Homeric Epics. Horn notes there are over 240 injuries in the *Iliad* and dozens in the *Odyssey* (almost all in Book 22). Three important points he makes are: (1) despite the constraints of the dactylic hexameter the typical Homeric battle between two combatants last no longer than a couple of verses. (2) The overwhelming number of such encounters end with a fatality. (3) Typically only major heroes such as Menelaus, Agamemnon, Diomedes, Odysseus, and, interestingly, the surgeon Machaon on the Greek side and Hector and Sarpedon on the Trojan side are afforded the possibility of a nonlethal injury.

In Chapters 13, through 9 Tyler Lucas, Anusha Lekshminarayanan and I examined the injuries in the *Iliad* to the head and neck, pelvis and lower extremity, abdomen, thorax, and upper extremity. In agreement with the classicists qualitatively, but with perhaps nontrivial quantitative differences we find that (1) injuries are often described succinctly, however, some engagements are more extended (see Tables 9.1, 10.1, 10.2, 11.1, 12.1, 13.1 and 13.2). (2) The fatality rate is extremely high (~85%+) but a battle injury in Homer does not mean certain death. (3) While almost always it is major heroes (and gods and goddesses) who sustain nonfatal injuries, there are exceptions, for example, Greek warrior Teucer and the Lykian Trojan ally Glaukis (see Table 9.1) suffer nonlethal injuries.

Horn further notes, and we find also, that not only are the injuries in the *Iliad* usually fatal, typically death comes very quickly to the wounded warrior. However, while Horn sees rapid death as stylization used by Homer giving status to the victor over the vanquished, we find that, constraints of dactylic hexameter notwithstanding, when a hero dies rapidly the anatomic description of the injury is always correctly one which will kill quickly, for example, fatal blows to the head and neck (Chapter 13), severing of a major artery (Chapters 9, 10, and 12), or significant injuries to important internal organs (Chapters 10 and 11). Conversely moreover, but to the same point, in cases where today we would think that death should not follow suddenly after an injury, for example, case 12 in Table 12.1 (see also entries 6 and 7 in Table 24.1) Homer gives a protracted course. Finally, in the *sui generis* case of a

Table 24.1 Illustrative cases from the *Iliad*.

Example number	*Iliad* ref.	*Iliad* text	Analysis	Comment
1	Book 4, lines 193–4	Therewith he [Agamemnon] spake to Talthybius, the godlike herald: "Talthybius, make haste to call hither Machaon, son of Asclepius, the peerless physician"	"Stat" (urgent) consult for Machaon to treat Menelaus.	This is the same protocol we follow today in (*organized*) medicine for an urgent consult.
2	Book 5, lines 290–295	[290] So spake he [Diomedes] and hurled; and Athene guided the spear upon his [Pandarus'] nose beside the eye, and it pierced through his white teeth. So the stubborn bronze shore off his tongue at its root, and the spear-point came out by the base of the chin. Then he fell from out the car, [295] and his armor all bright and flashing clanged upon him.		To have been able to describe the injury in such detail—the spear entering at the nose, then going through the teeth, then the tongue before finally coming out at the chin—Homer must have observed the injury up close.
3	Book 5, lines 405–440	burning corpses		Homer must have been deployed with an army to have seen this
4	Book 11, lines 251–255	and stabbed him [Agamemnon] full upon the arm below the elbow, and clean through went the point of the shining spear. Thereat shuddered Agamemnon king of men, [255] yet even so he ceased not from battle and war, but, wind-nurtured1 spear in hand, leapt upon Coön.	See case 86, Table 6.1.	Agamemnon, though stabbed, did not stop fighting. This must have been from an adrenaline rush. Wounded warriors would have reported this phenomenon to military physicians (or the military physician might experience this first hand after being wounded himself as Machaon was).

(Continued)

Table 24.1 (Continued)

Example number	*Iliad* ref.	*Iliad* text	Analysis	Comment
5	Book 11, lines 264–274	But Agamemnon ranged along the ranks of the other warriors [265] with spear and sword and great stones, so long as the blood welled yet warm from his wound. But when the wound waxed dry, and the blood ceased to flow, then sharp pains came upon the mighty son of Atreus. And even as when the sharp dart striketh a woman in travail, [270] the piercing dart that the Eilithyiae, the goddesses of childbirth, send—even the daughters of Hera that have in their keeping bitter pangs; even so sharp pains came upon the mighty son of Atreus. Then he leapt upon his chariot and bade his charioteer drive to the hollow ships, for he was sore pained at heart.	Once the adrenaline rush fades (see prior example above) Agamemnon compares the pain in his arm to the pain of giving birth.	Shows: a-Homer must have been there to hear the complaint. b- As a male he must have been a physician to know re childbirth. *c-Even back then they were using childbirth to norm the pain scale!
6,7	Book 5, lines 59–68 Book 13, lines 643–659	Injuries entering in the buttock and then piercing the urinary bladder.	Case 4, Table 12.1 Case 12, Table 12.1 See also Fig. 12.5 and Note 3 in Chapter 12.	The injury in Book 5 was by a spear and caused immediate death. The injury in Book 13 by an arrow and death was delayed. Homer must have been in theater to have observed the difference in the time course of death in the two cases.

8	Book 8, lines 320–329	[320] And himself Hector leapt to the ground from his gleaming car crying a terrible cry, and seizing a stone in his hand made right at Teucer, and his heart bade him smite him. Now Teucer had drawn forth from the quiver a bitter arrow, and laid it upon the string, but even as he was drawing it back Hector of the flashing helm [325] smote him beside the shoulder where the collar-bone parts the neck and the breast, where is the deadliest spot; even there as he aimed eagerly against him he smote him with the jagged stone, and he brake the bow-string; but his hand grew numb at the wrist, and he sank upon his knees and thus abode, and the bow fell from his hand.	First description of brachial plexus injury. See Fig. 15.1.	Homer must have seen this injury to be able to give such a correct description both in terms of location and time course of the injury.
9	13; 434–440	this Alcathous did Poseidon subdue beneath Idomeneus, [435] for he cast a spell upon his bright eyes and ensnared his glorious limbs that he might nowise flee backwards nor avoid the spear; but as he stood fixed, even as a pillar or a tree, high and leafy, the warrior Idomeneus smote him with a thrust of his spear full upon the breast, [440] and clave his coat of bronze round about him, that		Homer must have seen this injury—or gotten an immediate report from the battlefield—to describe it in such detail.

(*Continued*)

Table 24.1 (Continued)

Example number	*Iliad* ref.	*Iliad* text	Analysis	Comment
		aforetime ever warded death from his body, but now it rang harshly as it was cloven about the spear. And he fell with a thud, and the spear was fixed in his heart, that still beating made the butt thereof to quiver; howbeit, there at length did mighty Ares stay its fury.		
10	Book 14, lines 493–496	Him [Ilioneus (a Trojan)] then did Peneleos [a Greek] smite beneath the brow at the roots of the eyes, and drave out the eyeball, and the shaft went clean through the eye [495] and through the nape of the neck, and he sank down stretching out both his hands. [497 But Peneleos drawing his sharp sword let drive full upon his neck, and smote off to the ground the head with the helmet, and still the mighty spear stood in the eye]	Decerebrate positioning. See Chapter 15 for discussion of the case.	Predates Nobel Prize winner Charles Sherrington's description (Sherrington, 1898) by three millennia! Given the time course of events (see Chapter 15) Homer must have been present to see and observe the decerebrate positioning as a sequela of the head injury.

Author's own work. Translations (Homer, 1924).

nonanatomic recovery course—Hector's miraculously speedy recovery from a very severe chest contusion (see Note 2, Chapter 10)—Homer lets us know that he knows the course is nonmedical in a remarkable way by having a warrior speak this fact when (Book 15, lines 286–288) Thoas says, "Now look you, verily a great marvel is this that mine eyes behold, how that now he is risen again and hath avoided the fates, even Hector."

For many of the injuries classicists (see Saunders (1999) and refs. therein) had found to be problematic—that is, no plausible medical explanation (and also there not being one in the medical literature)—we have been able to provide straightforward anatomic solutions: see, for example, case 20 in Chapter 9, cases 3 and 5 in Chapter 12, and case 21 in Chapter 13. There are two injuries in Homer for which Horn finds the descriptions phantasmic or a melodramatic exaggeration. This may be the case or through unfortunate experience in a future war a clear anatomic or medical explanation may be found.

Homer's anatomic accuracy of at least more than 99.1% shows that he is not writing fiction when it comes to battle injuries and is strongly consistent with the idea that he attended to trauma casualties during a major campaign.

In Chapter 8, orthopedic surgeon Christos Koutserimpas discussed the management and treatment of trauma in Homer. Chapters 7 and 6, also by Koutserimpas, contain trauma registries for the *Odyssey* and the *Iliad*, respectively.

In Chapter 5, the last chapter of the **Introductory section** of the book, I discussed translations and editions of Homer. Have you started learning Homeric Greek? If not, time to start and experience Homer for yourself! In Chapter 4, I discussed the unique patronymic less, stateless, misshapen character Thersites and his exquisitely rare medical condition—cleidocranial dysostosis (CCD). You can't dream up CCD. Homer must have seen someone with the condition, or did in fact he himself have CCD (Altschuler, 2001)?! He seems to have an interest and fascination with the collar bone—see Tables 9.1, 10.1,and 10.2. Perhaps the wonder of one who did not have collar bones? Homer's knowledge of the anatomy around the collar bone is impressive—for example, the first description of a brachial plexus injury (see Figure 15.1). In Chapter 3, Christos Koutserimpas introduced the physicians and other medical personnel in the *Iliad* and the *Odyssey*. In Chapter 2, I went over Beasley's important, informative, and intriguing article (Beasley, 1972) about orthopedics in Homer which really got interest about medicine in Homer started in the twentieth century.

In what I believe is his last paper (Beasley, 2011) Beasley goes back to the topic of the collarbone or clavicle. He notes that its name derives from the Latin work *clavis* (key) in reference to its shape (see Fig. 24.1).

After discussing the clavicle's place in Vesalius' *De humani corporis fabrica libri septem* (1543) (Vesalius, 1543) and cases he operated on involving the clavicle over the years, Beasley ends by returning (Beasley, 1972) to Thersites a character born with no clavicles (cleidocranial dysostosis). Beasley asks how an individual with cleidocranial dysostosis could have gotten all the way with an army that was besieging Troy. He wonders if the humble clavicle could be a key to understanding the great adventure story of the *Iliad*. Could indeed the clavicle be the key to understanding Homer himself?!

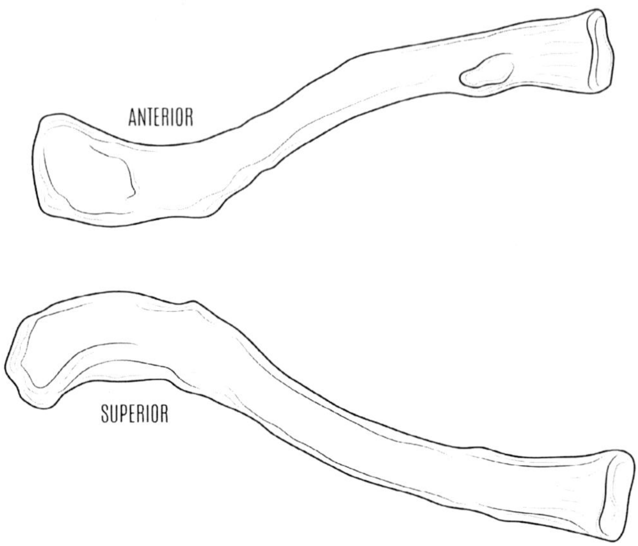

Figure 24.1 Drawing of clavicle. Anterior and superior views of the clavicle.
Source: Image by Mileidy Fernandez.

So, having learned so much about medicine in Homer, we return to where we started in Chapter 1: Who was Homer? A parsimonious reading of the facts presented herein suggests that Homer was a junior full-fledged physician or one at the end of his training, possibly with cleidocranial dysostosis himself who had deployed with a Greek army as a surgeon. This hypothesis is in accordance with Nobel Prize-winning physicist Richard Feynman's (1918–1988) comment about "the simplicity we typically associate with truth."

The sheer number of injuries in the *Iliad*, their brutality, and the often detailed descriptions are potentially themselves sufficient evidence in favor of Homer being an army surgeon who had experienced battle first hand (Melchionda, 1963). The time course and circumstances of some of the injuries and their descriptions, see Table 24.1, make it highly likely that Homer witnessed the injuries himself rather than hearing about them at a wine bar in Crete or elsewhere.[1]

Further to the notion that Homer was a military physician, Melchionda (1964) noted the importance of sports and sports medicine to the military physician. Indeed, both Epics contain sporting matches and events and sports-related injuries: Book 23 of the *Iliad* features funeral games overseen by Achilles in honor of Patroculus

[1] Interestingly, in a recent physiologic study Flouris and colleagues (https://journals.plos.org/plosone/article?id=10.1371/journal.pone.0301494) showed that modern special forces soldiers when wearing a replica of ancient Greek full body armor (which had been discovered in 1960 near the southern Greek village of Dendra) were able to complete an 11-hour day of combat maneuvers described in Homer's Iliad.

including chariot races and boxing matches. (For injuries during the games see cases 230–233 in Table 6.1.) In the *Odyssey*, there is a match-fight between the beggar Irus and Odysseus (in disguise as a beggar) on Ithaca set up by suitor Antinous for the amusement of the other suitors. (*Odyssey*, Book 18, lines 77–110. See cases 14 and 15 in Table 7.1 for injuries sustained during the fight.)

By comparison, Homer's discussion of geologic matters such as soil types, features of bodies of water land masses, and processes such as landslides, while specific and consistent (Zekkos *et al.*, 2008) shows knowledge, and crucially interest, at a lay level, not the specialist knowledge and passion he has for anatomy and trauma medicine.

How did Homer acquire all of his knowledge of anatomy in a time before MRIs, CT scans, and of course the internet? There are no known medical books or texts in Greek dating from Homer's time or before. Possibly Homer could have learned some from Egyptian medical papyri, for example, what is now known as the Edwin Smith Papyrus ((*Edwin Smith Papyrus*, 2024) named after an American antiques dealer who bought it in 1862). The Edwin Smith Papyrus, dating from Dynasties 16–17 of the Second Intermediate Period in c.1600 BCE is a surgical text written in Egyptian hieroglyphs (*Edwin Smith Papyrus*, 2024). The Edwin Smith Papyrus covers the diagnosis and treatment of 48 cases (Edwin Smith Papyrus, 2024; The Edwin Smith Surgical Papyrus Volume III, 2019) including head and neck injuries, clavicle dislocation (case 34), clavicle fracture (case 35) and humerus fractures (cases 36–38). Homer could also have learned from Babylonian medical tablets if these were available and he was able to read Assyrian (or Akkadian or Sumerian). The Epics do not belie any knowledge on Homer's part of any of these languages.

Most likely Homer learned surgery the way trainees do this day as an apprentice to an older more experienced surgeon. As a junior surgeon deployed with an army, would have, of necessity, provided Homer much experience from treating the many wounded. Further, he would have had the tragic windfall of a great many casualties to examine postmortem. (Homer's mom may even appear in the *Iliad* (see Box 24.1)!)

Not inconsistent with the *Iliad* having been written by a physician is the Greek warrior Idomeneus' comment to Nestor to help expedite care for the surgeon Machaon when he is injured "for a doctor is a man worth many others" (Book 11, line 514). Finally, the "economic argument" also favors the idea that Homer was a physician: Who paid for the Epics? They certainly do not seem to favor or support any particular King or city-state. The simplest explanation is that Homer supported composition and production of the *Iliad* and *Odyssey* via his surgical practice.

But we still have the same problem we started with: There is no other reference or way to confirm this hypothesis. We must then give Homer the last word, and indeed maybe Homer himself has confirmed things for us:

One of the great English madrigalist Thomas Weelkes's (1576?-1623) (somehow) Shakespearean anonymous text authors (Altschuler & Jansen, 2002) compares himself to Amphion (an ancient Greek mythological singer), Arion (a famed ancient Greek poet) and Phemius (a singer of tales from Ithaca). See Box 24.2 for the text of Weelkes's fantastic two-part madrigal (for six voices) "What have the Gods/Me thinks I hear" from Weelkes's 1600 set of madrigals in for 5 and 6 voices (What Have the Gods Consort of Musicke Anthony Rooley, 2013; Weelkes, 1600).

Box 24.1 Homer'smom.

Near the very end of Book 12 of the *Iliad*—the exact middle of the Epic—there may be a sighting (?!shoutout from a grateful son to his single mom) of Homer's mother! lines 430–435

πάντῃ δὴ πύργοι καὶ ἐπάλξιες αἵματι φωτῶν
ἐρράδατ' ἀμφοτέρωθεν ἀπὸ Τρώων καὶ Ἀχαιῶν.
ἀλλ' οὐδ' ὣς ἐδύναντο φόβον ποιῆσαι Ἀχαιῶν,
ἀλλ' ἔχον ὥς τε τάλαντα γυνὴ χερνῆτις ἀληθής,
ἥ τε σταθμὸν ἔχουσα καὶ εἴριον ἀμφὶς ἀνέλκει
ἰσάζουσ', ἵνα παισὶν ἀεικέα μισθὸν ἄρηται:

[430] Yea, everywhere the walls and battlements were spattered with blood of men from both sides, from Trojans and Achaeams alike. Howbeit even so they could not put the Achaeans to rout, but they held their ground, **as a careful woman that laboureth with her hands at spinning, holdeth the balance and raiseth the weight and the wool in either scale, making them equal, [435] that she may win a meagre wage for her children**; so evenly was strained their war and battle

This simile is rather unique in Homer in a number of ways: (1) the protagonist in the second part of the simile, the weaver woman, is a commoner. Homer almost never discusses or certainly does not feature commoners. (Thersites is the one male exception to this. See Chapter 4.) (2) In his similes Homer usually compares the actions of gods and men, to natural or pastoral scenes as in the simile from *Iliad*, Book 15, lines 173–175, not to affairs of daily life.

ἀγχοῦ δ' ἱσταμένη προσέφη κλυτὸν ἐννοσίγαιον:
'ἀγγελίην τινά τοι γαιήοχε κυανοχαῖτα
ἦλθον δεῦρο φέρουσα παραὶ Διὸς αἰγιόχοιο.

So [Zeus] spake he, and wind-footed, swift Iris failed not to hearken, but went down from the hills of Ida to sacred Ilios. [170] **And as when from the clouds there flieth snow or chill hail, driven by the blast of the North Wind that is born in the bright heaven**, even so fleetly sped in her eagerness swift Iris; and she drew nigh, and spake to the glorious Shaker of Earth [Poseidon]

(3) The comparison of the stalemate of the Trojan and Greek forces is not exactly the same as the equal balance maintained by the weaver on the loom: The Trojan and Greek forces are equally balanced, but these are *large* forces. Whereas, the balanced forces on the loom are *small* forces. So the simile is a force transducer or more exactly an "equilibrium transducer."

Amazingly, very near the end of the *Iliad* the mysterious (or not so mysterious) weaver makes another appearance[4]!

Book 23, lines 760–763

[4] In case, perhaps, one only had the scrolls for the second half of the Iliad and Homer wants to make sure the acknowledgment is in that part also.

[760] ἄγχι μάλ᾽, ὡς ὅτε τίς τε γυναικὸς ἐϋζώνοιο
στήθεός ἐστι κανών, ὅν τ᾽ εὖ μάλα χερσὶ τανύσσῃ
πηνίον ἐξέλκουσα παρὲκ μίτον, ἀγχόθι δ᾽ ἴσχει
στήθεος: ὧς Ὀδυσεὺς θέεν ἐγγύθεν, αὐτὰρ ὄπισθεν

Then speedily the son of Oïleus forged to the front, and close after him sped goodly Odysseus; [760] **close as is the weaving-rod to the breast of a fair-girdled woman, when she deftly draweth it in her hands, pulling the spool past the warp, and holdeth the rod nigh to her breast**; even so close behind ran Odysseus, [765] and his feet trod in the footsteps of Aias

The ever intrepid, totally dedicated, effective single mother, fascinatingly, is a common theme in the *Iliad*[5]: (1) Thetis arranging things with Zeus to assist her dishonored son Achilles (see Box. 5.1). (2) Aphrodite saves her son Aeneas' life after his pelvic/acetabular injury (case 4 in Table 12.1 and Note 4 in Chapter 12). (3) Hera raising her son Hephaestus to have a productive and important trade despite his congenital orthopedic deformities (Chapter 2). (Achilles' father the warrior Peleus king of Phthia is often mentioned in the *Iliad* but doesn't seem to have played a large role in child rearing, similarly for Aeneas' father the Trojan prince Anchises. Interestingly, while often in Greek mythology Zeus is listed as Hephaestus' father (*Hephaestus*, 2024), in the *Iliad* Hera is his only parent mentioned.[6] Also, within the texts of the Epics all of these sons are the only child of the union which produced them.)

In the *Odyssey* Odysseus' wife Penelope is a single mother to their son Telemachus for the twenty years that Odysseus is away wandering for ten years on his way home after fighting the Trojan War for ten years. Intriguingly, Penelope buys time staving off the suitors by weaving and then unweaving a garment.[7]

Author's own work.

[5] Interestingly, this is not inconsistent with Simone Swartzentruber-Emmons' point in Chapter 21 that Homer has women and men assume traditional gender roles.

[6] I believe we will never find Homer's father in the Iliad. If somehow Homer, like Thersites, had cleidocranial dysostosis his father likely took off leaving Homer's mother to raise him herself. This notion is consistent with Thersites being the only character in the Iliad without a patronymic. Indeed, even "bastard fathers" are fastidiously always listed/accounted for in the Epic.

[7] The paternal side not being completely left out, particularly notable is the long winded (what else?!) but detailed and specific instructions (Iliad Book, 23, lines 306–348) that Nestor gives to his son Antilochus before the chariot race at the funeral games Achilles has organized for Patroclus. What father is not like the Horseman of Gerenia—as Nestor is often referred to by Homer—in wanting his son to go into the same profession as himself? The fact that Homer informs of the detailed descriptions of instructions and hints on how to win the chariot race that were given to competitors (and also Homer's recounting of the bantering between and amongst spectators on the progress of the competitors) is consistent with him having been deployed with an army which had a chariot race and Homer having heard said descriptions being given. Had Homer not been deployed with an army but merely heard about the chariot race at some wine bar in Crete or elsewhere, he at most would have gotten reports of the final results of the race but most surely not the prerace instructions and the chatter at the course around these.

Box 24.2 What have the Gods/Me thinks I hear.

What have the Gods, the Gods their consort sent from heav'n
To charm my senses with heav'ns harmony
Care they for me, of all my joys bereaven
Send they heav'n's choir, to make me melody
Blessing me, with Music's felicity?
If it be so, great may your godheads be
And greater still to ease my misery

Methinks I hear Amphion's warbling strings
Arion's harp, distilling silv'ring sound
Orpheus' mean Lute with all in order brings
And with soul pleasing Music doth abound
Whilst that old Phemius softly plays the ground
O sweet consort, great may your comfort be
And greater still, to ease my misery

What did Homer think about Phemius? The age-old dictum in medicine states, "when in doubt, redo the history and physical exam." On history what do we find Homer thought Phemius was a collaborator who performed for the suitors trying to take Odysseus' wife and palace while Odysseus was away (*Odyssey*, Book 1, lines 325–327).[2]

What about the physical examination? Well, in this case it is an examination of Odysseus himself! When Odysseus returned to Ithaca disguised as a beggar he eventually was bathed by the nurse Eurycleia who had been Odysseus' nurse when he was a child. While bathing the {putative} beggar (*Odyssey*, Book 19, lines 465–474) Eurycleia notices a scar on his knee and immediately recognizes this as the one Odysseus had sustained while hunting wild boars with his grandfather Autolycus (*Odyssey*, Book 19, lines 449–451; see case 17 in Table 7.1).[3] (Nurses, like physicians, do their own assessment and physical examinations for new patient admissions/intakes.) Eurycleia joyfully identifies the beggar as Odysseus then swears herself to secrecy and assists Odysseus in killing the suitors, regaining his palace, and reuniting with Penelope.

So, Homer finds Phemius the singer to be a collaborator and the nurse a heroine. It sure seems likely that Homer was a surgeon who had been embedded with some Greek army and on return he wrote the *Iliad* and *Odyssey*.

[2] Later in the Odyssey after the suitors had been killed, Phemius begs Odysseus for his life claiming he did not play voluntarily, "god has planted in my heart all manner of lays" (Odyssey, Book 23, lines 347–8). Odysseus relents after Telemachus vouches for him.

[3] So important is this moment that "Odysseus' Scar" is the title of the first chapter of Erich Auerbach's classic work of literary theory Mimesis: The Representation of Reality in Western Literature (1946 in German, Verlag; 1953, Princeton University Press) (Auerbach, E. (1953)).

References

Altschuler, E. L. (2001). Cleidocranial dysostosis and the unity of the Homeric epics: An essay. *Clinical Orthopaedics and Related Research, 383*, 286–289.

Altschuler, E. L., & Jansen, W. (2002). Thomas Weelkes's text authors: Men of letters. *Musical Times, 143*(1879), 17–24. https://doi.org/10.2307/1004594.

Beasley, A. W. (1972). Homer and orthopaedics. *Clinical Orthopaedics and Related Research, 89*, 10–16.

Beasley, A. W. (2011). Weasels, keys and Thersites. *ANZ Journal of Surgery, 81*(12), 915–919. https://doi.org/10.1111/j.1445-2197.2011.05711.x.

Edwin Smith Papyrus. (2024). https://en.wikipedia.org/wiki/Edwin_Smith_Papyrus.

Hephaestus. (2024). https://en.wikipedia.org/wiki/Hephaestus#:~:text=H%E1%B8%97phaistos)%20is%20the%20Greek%20god,he%20was%20Hera%27s%20parthenogenous%20child.

Homer (1924). *The Iliad*. Harvard University Press.

Homer. (2007). The Odyssey of Homer.

Melchionda, E. (1963). Homer, a military physician? (I: The Iliad). *Annali di Medicina Navale, 68*.

Melchionda, E. (1964). Homer, a military physician? (I. The Iliad). *Annali di Medicina Navale, 69*.

Miranda, P. J., Baptista, M. S., & de Souza Pinto, S. E. (2018). The Odyssey's mythological network. *PLoS One, 13*(7). https://doi.org/10.1371/journal.pone.0200703.

Saunders, K. B. (1999). The wounds in "Iliad" 13-16. *Classical Quarterly, 49*(2), 345–363.

Sherrington, C. S. (1898). Decerebrate rigidity, and reflex coordination of movements. *The Journal of Physiology, 22*(4), 319–332.

The Edwin Smith Surgical Papyrus Volume III. (2019). https://archive.org/details/theedwinsmithsurgicalpapyrusvolumeiii/mode/2up.

Vesalius, A. (1543). De humani corporis fabrica libri septem. http://www.bvh.univ-tours.fr/Consult/consult.asp?numtable=B372615206%5F47294&numfiche=56&mode=3&ecran=0&offset=0.

Weelkes, T. (1600). What have the Gods/Methinks I hear. http://cpdl.org/wiki/images/9/96/Weelkes-WhatHaveTheGods.pdf.

What Have the Gods Consort of Musicke Anthony Rooley. (2013). https://www.youtube.com/watch?v=a9dAlu-k914.

Zekkos, D., Athanasopoulos, G., Zekkos, A. A., & Manousakis, I. (2008). Elements of engineering geology and geotechnical engineering in the Homeric poems. *History of Mechanism and Machine Science, 6*, 233–242. http://www.springer.com/series/7481.

Index

Note: Page numbers followed by "*f*," "*t*," and "*b*" refer to figures, tables, and boxes, respectively.

Made in United States
North Haven, CT
29 July 2025

71146912R00157